Storage

Storage

MERYL LLOYD & JOANNA COPESTICK

with photography by

ANDREW WOOD

TIME® LIFE BOOKS

Alexandria, Virginia

TIME® LIFE BOOKS

Time-Life Books is a division of Time Life Inc.
TIME LIFE INC.
President and CEO: George Artandi
TIME-LIFE BOOKS
President: Stephen R. Frary

TIME-LIFE CUSTOM PUBLISHING
Vice President and Publisher   Terry Newell
Vice President of Sales and Marketing   Neil Levin
Project Manager   Jennie Halfant
Director of Acquisitions   Jennifer Pearce
Director of Design   Christopher M. Register
Director of Special Markets   Liz Ziehl

Time-Life is a trademark of Time Warner Inc. U.S.A.

Books produced by Time-Life Custom Publishing are
available at a special bulk discount for promotional and
premium use. Custom adaptations can also be created to
meet your specific marketing goals. Call 1-800-323-5255.

Library of Congress Cataloging-in-Publication Data
Copestick, Joanna
      Storage / Joanna Copestick & Meryl Lloyd : with
photography by Andrew Wood.
      p. cm. -- (Essential style guides)
      ISBN 0-7370-0018-X
      1. Storage in the home. I. Lloyd, Meryl. 1958- . II.
Title. III. Series.
TX309.C67 1998
648'.8--dc21

First Published in Great Britain in 1998 by
Ryland Peters & Small, Cavendish House,
51–55 Mortimer Street, London W1N 7TD

Contents

Introduction

There was a time when it was exceedingly difficult to get excited about storage, but all that has changed as space becomes more precious, and belongings and technology accumulate at an alarming rate. Storage has become one of the most flexible and inventive factors of home planning. Sorting furniture and belongings and keeping them ordered and accessible has become a spiritual quest for the contemporary home owner. Where once was dull necessity, there is now a creative pursuit, a challenge that has to be met and needs that have to be resolved.

Storage is often a good starting point for designing a home from scratch. Thinking about your possessions, present and anticipated, is a good way of figuring out what your real needs are in terms of space, light, and furniture in your living space. To make your storage solutions work, you have to take a careful look at the kind of person you are. Someone who throws everything on the nearest available surface could benefit from a disciplined structure, where a labeled place for everything would offset a lackadaisical approach. A more ordered personality probably already has a sense of where things go and would allocate objects to natural homes close to their place of use.

As we accumulate more and more possessions, technology, and information, there is an increasing need for simpler homes that provide a calm space conducive to relaxation, in which to unwind and escape the telephone, fax, and doorbell. Organized living has become paramount as a way of controlling the work pressures, time demands, and fast pace of life that better communications and travel have brought to our everyday existence. Why sort through a pile of newspapers and children's drawings only to find a stack of unpaid bills and flyers for films and plays that have already come and gone? Good storage will ease the pressure and calm the mayhem: life is too short for chaos.

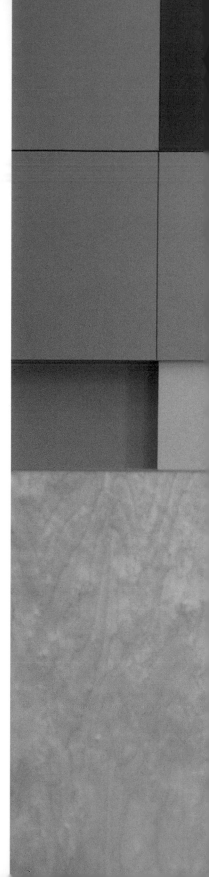

A way of life

Contemporary living makes so many demands on time, money, and space that, with our ever-increasing need to gather possessions around us, storage has become a buzz word for home owners. Good storage is held up as one way in which the quality of life can be improved, enabling you to create time for more rewarding pursuits.

Sorting out your storage needs and planning your living space accordingly is all a matter of working out a balanced view of how you live, how you want to live, and how you think you will be able to live. While a gloriously minimal kitchen with nothing on display may appeal to the control freak within, it is no use trying to impose an impossible order on a life that thrives on chaos. Yet if you feel as though you are constantly searching for mislaid items such as scissors or keys, perhaps it is time to examine ways to simplify your life.

The obvious starting point is to define how you live, then you can prioritize your storage needs according to your lifestyle. If you are an avid gardener, for instance, you will inevitably have numerous books and plenty of tools and paraphernalia, and if you have children, there will be ever-growing collections of toys, furniture, and leisure gear to contend with. Once you have carefully analyzed what you need to store, think about where best to store it. Will you need instant access to everything or can some items be put away in less accessible cupboards or out of the house altogether? Making your space work for you—every single corner of it—will let you relax in the area that is left.

Take a critical look at your existing storage space and decide whether it is adequate or whether you will need to make structural alterations or buy additional pieces of furniture to fit everything in. Perhaps you could make two rooms into one to create a larger space, and then devote a whole wall to built-in or freestanding

Below left Filing cabinets become storage drawers for hats, gloves, and shoes in an industrial-style cloakroom tucked behind sliding frosted-glass doors.

Top row left A container shaped like a giant hat box hides clutter and doubles as a bedside table.

Center row left Glass-lipped drawers are sleek, clean, and user-friendly, as their contents are always readily discernible.

Bottom row left Distressed almost to distraction, this closet has seen life. Pitted, worn surfaces are often more interesting than refined, clean-lined pieces that dare not hide real clutter.

Top row right Purpose-built shelves define a corner and enable toys and books to be stored on display in an accessible manner.

Center row right An Indian trunk shows that storage need not be obvious. Intricate carving and distressed paint effects render this box decorative as well as practical.

Bottom row right White shoe boxes tuck neatly into purpose-built slots that are generous enough to display shoes and their containers.

Baskets are versatile containers for everything from home-office clutter to toys and magazines.

This page **Serene order follows the strong, regular lines of this home office where even the stereo and television slip unobtrusively between neat wall units of drawers and shelving.**

Right top **Kitchens are sometimes best treated as angular planes of activity, the finished result being scrupulous order and calm.**

Right center **A bold-colored, geometric mini-hutch with a versatile combination of cupboards, shelves, and drawers prove that storage can be fun.**

Right bottom **Use unexpected containers, such as this battered case, to hold the unattractive necessities of modern living—audio tapes and CDs, or your worst bills.**

Far right **A giant-sized mirror salvaged from former glory makes a perfect centerpiece in a big, white hallway.**

units. Choosing between single pieces of furniture and built-in options depends a lot on the available space, but as a general rule, built-in cabinets and shelves take up less space and work more efficiently than stand-alone items, especially in kitchens.

Editing the contents of each room in an ordered manner—however disorganized you are by nature—is a necessary chore to start the process. The task will seem less daunting if you set a time limit and confine your efforts to one drawer, one closet, or one room at a time. Often a determination to clear a lot of clutter at once is thwarted as you find yourself dwelling on old diaries, photographs, and gifts from jilted lovers. Sort your possessions into piles, containing the emotion for as long as it takes to create the most satisfying pile of all, marked "out." Out of the closet and out of your life, to yield more space for everything else.

Left **Display collections
in a way that enhances
and emphasizes their finer
points. Period shoes are
perfectly aligned in almost
comical regimentation.**

Below **See-through
containers do not have to
be scrupulously organized.**

Right top **Neat shelves
recessed around a door are
perhaps the most space-
saving solution for any room.**

Right bottom
**Battered cabinets with
many compartments mean
that horders can hide their
furtive collections of shells,
stamps, or audio tapes.**

Far right **There will never
be fights with a pair
of sinks on generous
cupboards topped with
a long mirror.**

Categorize discarded clothes, china, unused presents, or worn-out furniture according to whether they should go to the thrift store, recycling center, or simply into the trash can. Garage or yard sales are a good way of off-loading unwanted objects—you can always go home via the dump to rid yourself of what's left over.

Create another pile for everything you need to keep but don't want cluttering up the floor of the closet or lurking in a drawer: photographs, letters, newspaper clippings, seasonal clothes and decorations, and the like. Buy boxes, baskets, or files, and label everything that goes into an attic, shed, garage, or cupboard. Store everything you decide to keep with other similar items: mixing up too many disparate objects is a sure way of losing them.

What you are left with should be only the necessary and the beautiful: those items that you need to store close at hand for frequent use and objects that demand to go on display, whether on walls or shelves, or as freestanding pieces.

Creating space

Making alterations to the structural elements of your home can often yield unexpected and fantastic results. Tearing out a false ceiling, for example, or making one really tall, airy room instead of two low ones will give dramatic impact to the space, as well as create extra height and much more wall space for hanging, displaying, and building into. Knocking through two smaller rooms to create one large space often produces more light, and tucking closets, rooms, and windows into the eaves will give you space you never knew existed before. It is all a matter of opening your mind to the less obvious solutions that your home presents, looking at a space with a sense of vision from all possible angles, and searching out ever-more ingenious ways to use whole rooms, rather than just a part of them, as effectively as possible.

While many architectural storage solutions are born of fascination with minimalism—a sense of order and calm often go hand in hand with a compulsion to contain belongings in strict regiments of flush-fitting cabinets—it doesn't have to be a lifestyle choice. One wall in a bedroom or living room with sleek floor-to-ceiling built-in cabinets may be all you need to hide a serious music collection, a library of children's videos, and five-years' worth of your favorite magazines. There is something undeniably appealing about a neat wall that gives the appearance of control and space while concealing a tawdry conglomeration of old batteries, broken toys, and spare plugs. For those who yearn for a pristine white space with no visible objects on display, building in some simple streamlined cupboards may be all that is required to help you achieve this. The thought of preserving the symmetry created by the clean lines may be enough to persuade you not to give in to a natural inclination to leave correspondence, books, and newspapers lying around—you may even find yourself inspired to keep their contents in order, too.

Opposite **Floor-to-ceiling control is what you need to achieve a successful approach to minimal living. The taller the cabinet, the more junk you can cram in.**

Left **In a cool, calm white space the recessed kitchen countertop is deliberately painted black so that, come evening, dirty plates can be piled up and not even seen under the subtle overhead lighting.**

Below **An unobtrusive under-counter unit opens to reveal order within. A row of inviting glasses in a clean, sparse cabinet is something to be proud of—a sense of order that is reflected in the outward appearance of this kitchen.**

It is obviously easier to incorporate architectural solutions during the early stages of planning or redesigning a space, but individual rooms can be altered one at a time, too. Remodeling a kitchen may give you a chance to break away from using more traditional materials and layouts and go for a bolder approach, while knocking two rooms into one gives you a chance to treat the whole of one side of the room in a different way, using low-level cupboards, integral shelving, or raised flooring with storage beneath. The materials you choose are important if you wish to match the storage with existing architectural features or fixtures.

Adding on rooms in the form of attic conversions, basements, garages, sheds, and sunrooms will provide additional living and storage areas. In small spaces, sliding doors, pull-down beds, cupboards that contain televisions or pull-out ironing boards, and closets with folding doors concealing storage on casters, are all good ways of making use of tight space. Take inspiration from the compact living spaces in houseboats and motor homes.

Left **When even the oven is tucked away behind a door, you know there is a serious minimalist at work. The message is clear—nothing but ultimate tidiness will do in this kitchen.**

This page **Plain cabinets embrace the different elements of a kitchen, a home entertainment area, and an extensive book and piano-music collection without anyone having a clue as to the contents of the room.**

Rooms for storage

An ordered environment has an undeniably calming effect, and if you are the kind of person who never manages to file the bills, find the spare buttons, or locate a new bulb, take a fresh look at your home room by room. We have divided the home into fluid, individual zones, where activities, relaxation, and access to your belongings is treated as a series of life cycles, instead of examining storage solutions in isolation—a holisitic approach. Think about the natural path belongings follow around the home and identify the best place to store them.

Many items, such as clothes, follow a circular track from closet to laundry basket to washer, requiring several storage solutions on route. Creating an easy path through these stages is as vital as allocating unobtrusive storage that merges into the surroundings with ease, making everything accessible and usable.

A place where

the living is easy

The pleasure center

Living rooms are different things to different people. In some homes a living space is the one room in the house in which the whole family congregates to relax, entertain, and play. In others, it is a more formal space into which guests are invited, and the television is banished to another room. Sometimes a living room doubles as a work space, an eating area, or a playroom. Figure out your priorities and what you use the room for most, then plan your storage accordingly so it works best for you.

A room in which you read, relax, and entertain guests is one space where large, freestanding cabinets make a lot of sense. Often, characterful old pieces, such as a battered hutch, an antique armoire, or even an old metal school locker, are perfect for concealing video tapes, audio cassettes, home entertainment centers, and the "to file" pile. Many people prefer to store at a least part of their book collection in the living room, since it is the place where you are most likely to find some spare time to start or finish reading a novel. If there are no convenient alcoves in which to fit shelving or cupboards, devote part or all of one wall to a line of deep, invisibly mounted shelving.

Pass a critical eye over your existing storage and furniture arrangements. The advantage of

having a house full of freestanding furniture means you are able to move pieces around as demands and tastes change. Glass-fronted cabinets, pigeon-hole shelf units, open shelves, and linen cupboards will all work equally well whether they are in a kitchen, a bedroom, a study, or a living room, giving you the freedom to move pieces from room to room as your needs change.

Modular systems are a convenient halfway house between freestanding units and built-in solutions. Take your pick from a huge variety of decorative styles and configurations to allow you to achieve exactly the right combination for your specific needs.

Whether your preference is for built-in or freestanding options, choose pieces of furniture with multipurpose uses.

In the same way that kitchens can be revamped by changing or painting the cupboard doors and replacing the door hardware, alcove cupboards in the living room of a recently purchased home can be similarly altered to suit your taste. In time, you may be able to have a carpenter remodel the space with your own design. If you do, start by finding pictures of similar systems in a magazine or book, make a rough sketch, and discuss your ideas with a carpenter. If you are not planning to stay in your home for more than a couple of years, it is probably not worth considering building in too many cabinets and closets. Although ample storage is often a good selling point, the chances are the new owners will want to impose their own decorative style on existing cupboards and shelving systems.

The most expensive items in a living room are usually the seating arrangements. If space is limited, you may prefer to have a sofa bed instead of a large armchair, or an ottoman that will double up as a side table or coffee table. Choose pieces of furniture that have

Home entertainment can become unmanageable, unsightly clutter without proper shelves, cupboards, or racks.

multipurpose uses. Low-level coffee tables that incorporate drawers will easily hide books, magazines, and candles, while nests of tables —a great space saver—are no longer considered passé. Gate-leg tables can transmute from small display areas when they are folded up to miniature desks for homework or bill-paying when extended.

Make sure your storage exists happily alongside the decoration and architectural style of your space without compromising the visual impact of the room or its relaxing function. Exploit the existing architecture, whether it is alcoves on each side of a fireplace, a tall,

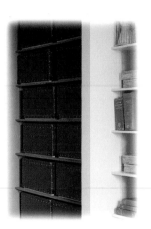

built-in corner cabinet, or a square-bayed window that lends itself to a window seat with under-bench storage. High-level shelves and Shaker-style peg rails make good display areas at a safe distance above rampaging children, while on the floor old trunks, blanket boxes, and lidded baskets are good movable options. Slim alcoves between large casement windows can be fitted with narrow shelves and cupboards, while wide doorways can be enhanced with bookshelves that fit around and above the door frame.

Don't forget the quirky solutions; they are often the best. Make use of redundant industrial furniture, such as utilitarian filing cabinets fitted with casters, battered old school lockers, and catering-style metal shelving. Old baker's trolleys make smart movable homes for ever-increasing piles of current magazines, books, or videos, while unused map chests are perfect for preserving a precious collection of children's artwork and paperwork.

Display is important in a living room, since it is the place where you are most likely to want to view and admire favorite collections and treasured objects. Good lighting is just as important for highlighting groups of pictures in a graphic arrangement as it is for spotlighting small objects in a wall-mounted glass display case. Uplighters will enhance any treasure placed along a picture rail, while a simple halogen table lamp will throw a pool of light onto favorite picture frames, glass vases, or ornaments.

Organically shaped frameworks will often elevate a repetitive collection of objects, such as old tin toys, baseball cards, or ornaments to a well-considered display piece. An array of similar-size pieces arranged and displayed together in a sensitive manner will make a pleasing, personal addition to any living space.

Create a stress-free living space that is conducive to relaxation.

Baskets make attractive, versatile, and useful storage options. Brightly colored flat-bottomed varieties with handles can be used to carry magazines and other reading matter around the room. Log baskets beside a fireplace are decorative and functional. Large box-shaped versions will house newspapers and books, while smaller ones are perfect for spare loose change, keys, and recent receipts. Small-scale storage is a good way of combining clutter-control with display: miniature chests of drawers, wooden boxes, and Shaker swallowtail oval boxes are all visually appealing yet practical.

Cool, calm, and under control

Left **Serenity is the only possible state of mind in a carefully considered living space where everyday junk has retreated behind sleek, regular cupboard doors.**

Center top **Large shallow drawers hold a serious tape collection, so each is visible.**

Center bottom **Hiding the stereo may not be everyone's ideal, but it maintains the lines of a pristine space.**

Right **There may be chaos concealed behind these measured doors.**

When a space is planned from the start with integral storage in mind, the architecture of the room merges easily with its contents, as each piece of furniture can be selected to work with the others to complement the natural contours of the room. Here, clean white lines and a rich mix of textures and materials combine to create visual peace in a remodeled house. The contemporary-style floor-to-ceiling cupboards and discreet benches that open to allow for more storage mean that everyday clutter is effortlessly hidden from view, allowing the soft lighting and a sense of control and restraint to come to the fore. The honest, hard-working qualities of the natural materials used—wood, stone, and leather—create a surface harmony that does away with a need for pattern or paraphernalia.

Words and pictures

Placing works of art and books so they are easy to reach yet pleasing to the eye is often a problem in living areas that demand to be relaxing retreats. A large collection of books can be visually disturbing and too dominant, but placing them in glass-fronted cabinets or behind sliding panels or swivel cupboard fronts helps contain their presence, while the dead space of cupboard doors can be used to display pictures. Filling cabinets below book shelves leaves the rest of the room free for furniture.

Cantilevered cupboard doors slide along grooves to serve as window shutters, mock walls, and display space—truly multifunctional.

Far left **A striking cherry wood ethnological display cabinet is also home to carved wooden clubs and ancient fossils.**

Center left **Plastic cubes double as side table and display cabinet without to taking up much space.**

Center above **A picture gallery is a cupboard door which opens to reveal books and more pictures.**

Open house

A functional approach to storing the things you need around you is a good way of turning necessities into decoration. In an industrial-inspired living area, open book shelves abut glass-fronted cupboards that are actually a pantry. Every space on the two walls has been used, leaving the floor clear for basic, well-considered furniture—like the long table that serves as both a work station and dining area. The shelves are on adjustable brackets to accommodate all shapes and sizes of books, boxes, and files, while under part of the bottom shelf are shallow drawers for tapes and stationery. The glass panels along the other wall create colorful reflections from the wall of books and the plants on the decking outside. Within the cupboards are china, glasses, food, and appliances, which are handily close to the kitchen area on the right, and to the table. The antithesis of a minimal aesthetic, yet contemporary in its approach, this is a modern Aladdin's cave.

Above **Open-plan living spaces demand that storage is practical and placed exactly in the area where it is needed. Here, the four walls of one huge room each has its own particular function.**

Right **Library meets office on a wall devoted to books and files. There is no excuse for not finding a particular file, book, or photograph within such a visible and comprehensive display of the written word.**

Sound and vision

Far left **Placing the television on open shelves among a large magazine collection integrates it very well and prevents it from becoming a dominant feature in the room.**

Left **Packing a stereo cabinet into a slatted cupboard is one way of de-teching the living room.**

Above **Small CD boxes stacked on industrial metal shelving make their own storage system.**

Above right **A stereo tucks neatly into a shelf space under a run of wall cupboards and is at a perfect height for easy use.**

Music and television are indispensable in a living space where you can kick off your shoes, forget the outside world, and learn to relax. The disguise-versus-display debate often rages here. Many music buffs would never dream of hiding cherished albums, CDs, or tapes in cupboards or boxes, whereas others might yearn to hear but never see them. Either way, the answer is to store the stereo close to the music collection, in the area where you most listen to music. Drawers in cabinets or on casters, shelves, alcove cupboards, and freestanding units are all possible solutions, but cabinets should allow the air to circulate, as stereo systems emit heat and need to breathe. Stereos sound better off the ground, as do speakers unless they are on integral stands. Integrated storage systems that house all the technology, including the TV, cable and satellite decoders, stereo, and even a computer are available in designs for every architectural period.

Balance, scale, and perspective

Opposite **Floor-to-ceiling cupboards and a concrete fireplace merge organically with the architectural knocked-through space.**

This page **Narrow geometric cupboards with sharp-angled handles extend all the way up to the ceiling to shield all manner of unwanted mess from view.**

Architect-designed homes often display a great respect for storage and what it can do to a space. Designing storage so it enhances a room, extends the architectural story, and fulfills the functional needs is a strong consideration in homes where specific belongings and attitudes push the design in a certain direction. Here, a collection of books and magazines had to be incorporated in a clean, pared-down space, where the supreme quality of every piece of furniture and fixture dictates the pace. In a neat recess, chunky, invisibly supported shelves are given importance but don't dominate. Above a stone fireplace is a thick-lipped mirror of the same width, and in the center of the room, a square table reflects the sharp angles of the slatted seats. All is symmetrical, balanced, and calm.

Above **A gnarled ladder lies along a shelf in a quirky acknowledgment of the sharp lines and right angles.**

Right **Smooth surfaces and strict lines are softened by a rug and banquette-style sofa.**

Boxed away and stacked up high

Above, left to right Wooden oval boxes make an elegant display that doubles as an end table; a versatile freestanding unit can be used for a variety of functions—side or bedside table, vegetable storage, or toy chest; a casual stack of colorful boxes looks attractive on display.

Far right Order is imposed by a collection of boxes that slot neatly into made-to-measure shelves.

In any living space, there will always be myriad small items that regularly need clearing up and containing, often at the end of a busy day or after a weekend of entertaining when you most need to surround yourself with order and establish a sense of calm. Freestanding furniture, baskets, boxes, and trunks in interesting shapes and a variety of sizes can hide everything from magazines and vacation pamphlets to photograph albums and oversized books.

Whatever you need to store, unexpected, unusual units with a hint of humor will disguise all sorts of uninspiring, unattractive items, as well as livening up the surroundings. For tiny objects, use old decorated

Different-sized units of the same design are graphic statements of excellent organization.

cake pans, large glass or pottery jars, galvanized buckets, or deep willow baskets. Concealed drawers, old tin trunks, and blanket boxes also look decorative as well as serving a practical function.

Rather than store small boxes on wall-mounted shelves, incorporate them into a display by piling them up high on the floor or stacking them against a sofa for a small, instant side table. Low-level coffee tables with integral drawers can hide a surprising quantity of clutter—even the remote control can claim a home—and provide another surface for display.

Mixing old and new

Opposite **Arched windows throw light on 1960s furniture and modular storage.**

Left **A multifunctional coffee table is an asset in a compact living space.**

Below **Matching bowls make decorative containers for small change or keys.**

Part of a large house that has been converted into a compact apartment combines contemporary design with traditional architectural strengths and features. Low-level cupboards tuck underneath lift-up shelves, with a slab of slate in the center to create a sleek hearth for a recessed fireplace that tips a hat to 1960s design. On either side, cubelike flush-fitted wall cupboards are interspersed with alcove shelves to form mini storage systems.

A sense of order is preserved with a low-level coffee table that has three pull-out drawers, a large wooden trunk, and several wicker baskets for magazines and books. In a small space, keeping clutter tucked behind cabinet doors and on the wall allows the rest of the room to be devoted to necessary furniture.

Far left **A makeshift vase in the form of a battered and rusting metal pitcher is perfectly at home on a rough-hewn mantelpiece.**

Left above **Pitchers and other small items, such as cups and saucers, candle-sticks, and boxes, create small displays interspersed with books on a shelf.**

Left below **A graphic checkerboard of built-in cupboards and shelves calls out for favorite pieces to be displayed interspersed among the book collection.**

Right **A delightful glass-fronted linen cupboard makes a versatile, multifunctional piece of furniture in a living space.**

Details on display

Finding the right home for your favorite vase, pitcher, bowl, or plant container, where it is always easy to locate and get at when you want it, is sometimes an organic process that evolves depending on how firmly your chosen piece demands to be on display. Another time, you may find yourself with a specific piece of built-in shelving that is crying out for a graphic display of disparate objects or collected treasures and would look empty and unfinished if it were not filled. Or, you may have a glass-fronted cabinet that needs filling with careful attention to detail and aesthetics, rather than simply cramming it with whatever comes to hand. Display that works can truly be an uplifting aspect of storage.

Energy store

Kitchens that

function with style

The kitchen is the one room in the home where fundamental storage requirements have to be addressed, but limitless solutions are on offer. The idea of the built-in kitchen first developed in the U.S. during the 1920s, where small pantries evolved into the built-in cabinets we find in kitchens everywhere today. They remain a potent influence on how we plan the hub of the home, complete with catalogs that ache with mind-numbing gadgets for keeping the vacuum-cleaner tubes straight, cups well hung, and spice jars firmly attached to cupboard doors. While some of these are undeniably useful and ingenious, others are just time wasters. The trick is to see past the gimmicks to the cool solutions.

During the 1990s the trend in kitchen design has been to veer away from the strict uniformity and constraint of lines of cabinets arranged around a room. Yet cupboards topped with work surfaces are a system that works well, and in a small kitchen built-in units still make the best use of space. They can be made more flexible by adding trolleys or shelf units on casters underneath the counter. If finances are tight, this is a good way of preserving some space for when you can stretch to a dishwasher. In the meantime, you can use the space for hiding the recycling bin or unsightly cleaning equipment.

To sort out the kitchen and get to grips with the storage issues, start by looking critically at all your kitchen equipment. Lay it out on the kitchen table, including every last piece of unused wedding-present china, seldom-seen baking trays, sad-looking casserole dishes, and discarded electric carving knives. Edit individual pieces with several criteria in mind. When was it last used? Is it past its "styled-by" date? Is it simple to use or do you always cut your fingers on it? Does it waste time rather than save time? Divide everything into

three piles for a yard sale, the dump, and hand-me-downs, and then store any sentimental heirlooms out of the way in the attic or cellar.

Then separate everything you have left into four categories. Think about each item and decide whether you need it readily accessible for everyday use: you must keep it on display because it is pleasing to look at or is always needed, you can happily relegate it to the back of a cupboard for use only twice a year, and, most importantly, you don't know why you've hung on to it for so long. This last pile embraces all the items you think you can't throw away yet. Put these at the back of a cupboard and make a note to check them again in three months' time. If you still haven't used them, they simply have to go.

Don't forget all the paperwork, too. Some say that if you keep only one drawer in the kitchen for miscellaneous clutter, then that is as much as you can accumulate. The more nooks and crannies you

Spring cleaning the cupboards on a regular basis helps keep the kitchen functional and under control.

devote to junk mail, spare batteries, and stationery, the more chance there is to overfill them with items that never see the light of day.

Once you have filled a few garbage sacks, you will probably have edited down the contents of your kitchen to a manageable level. This process is worth doing annually at least, even if you aren't planning to rip out the kitchen and start again. Judicious evaluating and restocking that leaves you with a clean, sorted cupboard full of good things is a satisfying chore that is always appealing. Learn to live with less; it will lighten your load, physically and spiritually.

Once edited, divide your belongings into functional groups. Cooking equipment and appliances in one corner, china in another, flatware and table linens in another. You will find that a natural order emerges. If you are planning a kitchen from scratch, decide how you are going to live in it. If there is an eating area as well, you may want to store some table linens and glasses close to that spot. If the space benefits from a high ceiling, a hanging rack could be

Make the most of any architectural features that will make planning kitchen storage an easier task.

an ideal space-saver. An existing fireplace alcove calls out for a cooking range, freestanding hutch or sideboard built to fit. Make sure you cover the range in scale, taking into account everything from your large saucepans and cumbersome food processors to tiny tea strainers, mustard spoons, and nutmeg graters. Small storage containers, large built-in or freestanding cupboards, pull-out pantries, and vegetable racks on casters, baskets to slot under surfaces, and wall-mounted racks for utensils are all simple solutions that make a kitchen work efficiently, aesthetically, and safely.

Next, figure out what existing storage there is and what can be added. In a kitchen where, despite getting rid of lots of items, storage space is still tight, look at the walls and see if there is any space left for additional shelving or cupboards. If there is floor space available, buy a butcher's block or trolley on casters for moving food, china, or homework from one area to another.

Food shopping is not undertaken as frequently as it once was. We tend to buy in bulk once a week or even less often, so more food needs to be

stored at any one time. Big fridges are one of life's necessities, and all perishable foods should be stored there. There is no sight more pleasing than a generously proportioned fridge, well organized and interestingly filled. For maximum life, fresh vegetables should be kept in a dark place where the air can circulate freely. Try to keep all your cooking necessities, such as olive oil, oriental sauces, condiments, and spices, close to the stove, and have legumes, pastas, and noodles displayed or in a cupboard, close at hand, so you don't have to keep walking from one area of the room to another while you're cooking.

Display is a strong element of storage in kitchens. Making a virtue of appliances, utensils, and equipment means throwing out chipped and ugly china and saucepans and replacing them, over time, with your favorite materials and makes. Stainless-steel and copper pans, pleasing toasters, coffee makers, and a gleaming muti-purpose food processor are beautiful objects in their own right. There is no need to hide them away behind cupboard doors if they are in constant use. China is another element that makes a great display while being easy to access. Display everyday pieces close to the dishwasher to make the life cycle work for you, not against you.

Keep stylishly designed appliances and small-scale storage solutions on display, freeing up cupboards for other items.

Fresh fruit, bottles of mineral water, and ample supplies of wine can be good to look at as well as useful, along with vases, big bowls, and baskets. The following pages show the sheer scope of workable, imaginative solutions for the energy store.

From store to store

Storing food is, in a sense, a matter of recycling. The process begins the moment you bring food home from the store. Individual items are categorized and put away, only soon to be removed, prepared, cooked and eaten and the waste discarded in a cycle as constant as life itself.

This page **A range of flush-fitted units slotted neatly beneath chunky marble countertops blend subtly with the white walls, providing a neutral back-drop for the kitchen table, where the shopping can be dumped, ready for sorting.**

Right above **Keep all of your fresh vegetables in a cool place, out of the light, and in a container that allows the air to circulate.**

Right center **Pasta always looks its best—and is easy to identify—stored in transparent containers.**

Right below **Wine bottles make a graphic display in the right kind of rack.**

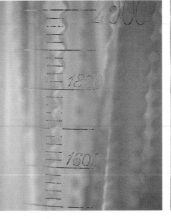

Above top **A big fridge eases the burden on storage elsewhere.**

Above left **Think laterally and use unexpected storage containers, such as glass** measuring beakers, which look cool filled up with long ribbons of pasta.

Above right **Sort fruits and nuts by shape in glass tanks for simple visual order.**

Freestanding

Strong, dominant colors create a powerful impact in this kitchen, where chunky, free-standing wooden units create a sense of the built-in. On an industrial metal shelving system against a sharp red backdrop, a visual drama is played out by an eclectic combination of pasta and legumes, cook books, a microwave, pots and pans.

The freestanding units are all designed in the same style, but each one has a unique function. A cooking zone centers around central burners with a

units for flexibility

counter on each side and shelves underneath, perfect for cooking trays and oversized bowls. The washing zone comprises sink, dishwasher, and cupboards which, lacquered in a shade of red that sings with vitality, shine above the stone floor. The central island is primarily for food preparation and has large pigeonholes in which everyday china and glasses are stacked ready for use, making the trip from the dishwasher opposite an efficient operation. On one side of the counter are small, square drawers that contain jars of herbs, teabags, and other clutter-inducing objects. The island encourages people to stand and linger against its bar-style countertop.

Left **Generous freestanding units are simple but ultra-effective, combining open shelving, concealed cupboards, pigeonholes, and drawers for versatile and flexible storage.**

Right **An island work station incorporates handy drawers for clutter. A useful, movable spice rack on the counter holds glass vials that can be filled up with herbs as necessary.**

Utilitarian style

This loft is home to a couple who have resisted the pull of retirement in the country in favor of a bold reappraisal of city living. Storage is both sensible and adventurous in a kitchen that takes up one corner of one level, where metal-rimmed windows allow light to bounce off a glass backsplash and gleaming stainless-steel accessories.

Chrome, used with discretion, is always capable of pulling a room in a contemporary direction without becoming cliché. Here, it slips neatly onto several surfaces for a unifying visual coherence. Teamed with white cabinets, tall food and pantry cupboards, and a polished wooden floor, the end result is a sleek mix of functional materials and no-nonsense accessories. Designing cabinets in such a way means they are at home in any room space, and, whether in the kitchen, living room, bedroom, or bathroom, they make a quiet backdrop to the serious elements of cooking, entertaining, and living.

Left **Heaven is a big blue fridge with a plinth all of its own to emphasize the fine contours and bold color.**

Right **Chunky recessed handles and oversized chrome pulls emphasize the practicality of this attic kitchen.**

Far right **A pull-out pantry is many people's idea of condiment contentment. Everything is at once visible and accessible. Ideal for throwing together a hasty snack.**

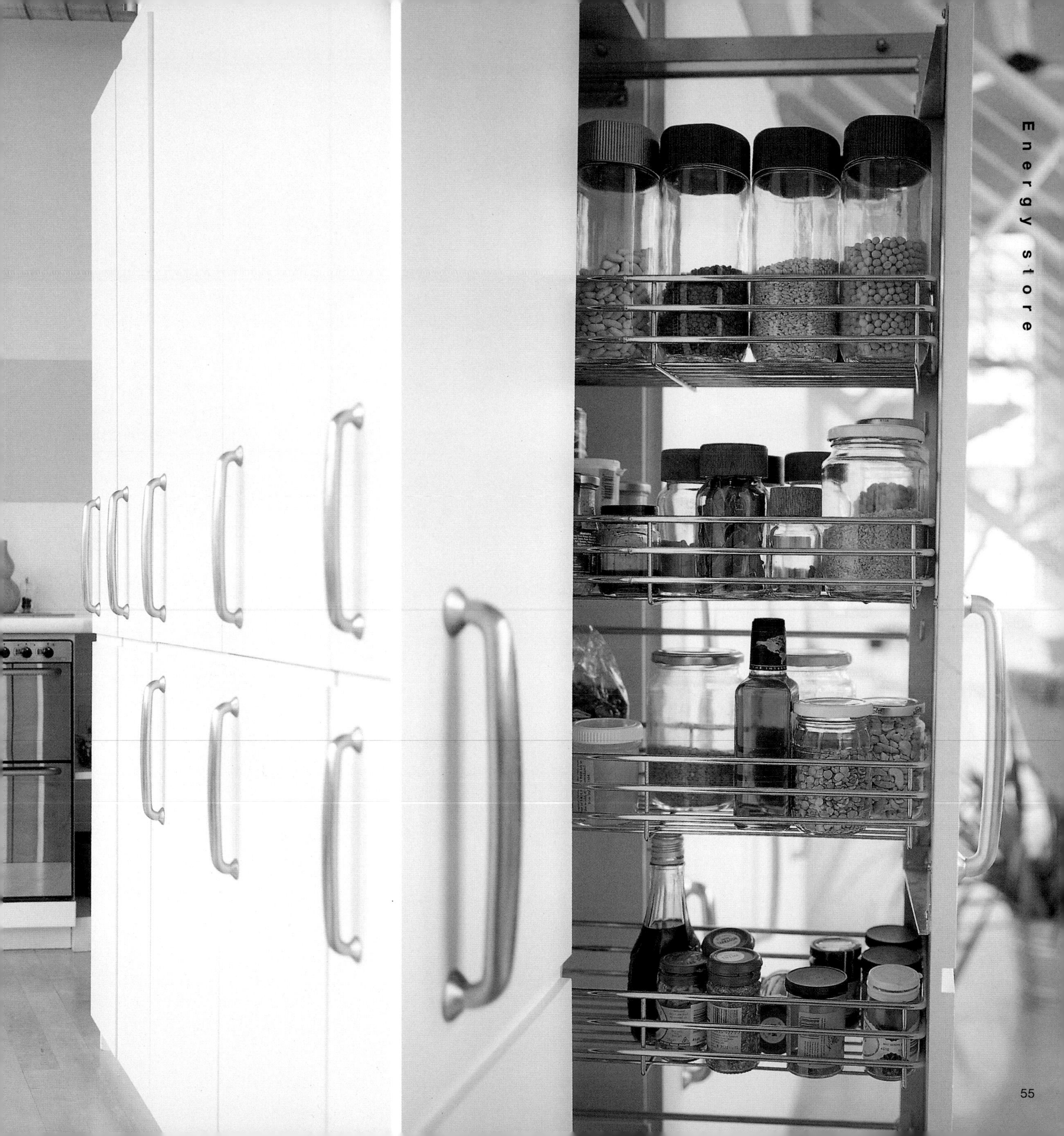

This page **Find battered kitchenalia in specialized antique stores or yard sales and use them to store unexpected items.**

Right above **Traditional bottle racks are constantly being reinterpreted in new materials, such as bright plastic, chrome, and steel.**

Right below **Transparent containers do away with labels, which eventually peel off or wear away.**

Far right **If space permits, bread bins are a great asset and can look good on the countertop.**

Far right above **A neat chrome trolley will store, transport, and display food, china, or table linens.**

Far right below **Funky metal canisters in a variety of sizes make ideal containers for tea, coffee, and sugar.**

On the shelf

I f you are not someone who likes to keep everything put away, feel free to go with your instincts and grasp the notion of big-time display. As long as there is space for everything you need in the kitchen, it doesn't matter that china, utensils, and pans mingle together and jostle for space in bazaarlike decoration.

In this converted schoolhouse, wicker picnic baskets hide a collection of uninspiring plastic containers and distinctly unstylish colanders and strainers, yet despite the relaxed, casual style, they look neat and controlled lined up in threes in a space beneath the countertop. On the other side of the room, a painted carved French armoire is used as the pantry and is kept permanently stocked up with cans of vegetables, powdered milk, and sauces for impromptu meals.

In a room where storage is so much a part of the decorative order, the addition of an extra-long table and informal chairs creates such a relaxed atmosphere that the absence of conventional solutions, such as built-in units, concealed cupboards, and individual cooking, eating, and entertaining zones, matters not a bit.

Left **Using an armoire or utility cupboard as a pantry is what everyone did before built-in kitchens were invented. The deep shelves in antique cupboards are a good place to stash key ingredients, such as bulky packets of pasta, oils, mustards, and vinegars.**

Right **Keeping all the kitchen equipment out on display means you can always locate what you need and never have to make more than a couple of steps to the nearest pot, pan, or utensil.**

Left **Pantries can be as relaxed in appearance as old cupboards or as sophisticated as pull-out units with individual storage baskets. As long as they are stocked up with good things and you know where to find everything you need, that's all that matters.**

Right **Supporting a tiled counter with wooden supports and shelves is an economic alternative to fitting out an entire room. Dot baskets within for keeping larger items out of sight, or else pile up plates, bowls, or vases to combine storage with display.**

Left above **Tucking a small fridge under a counter is a good way of preserving space. You can always have a large fridge–freezer elsewhere to top off supplies.**

Left below **Chopping boards stack neatly under a table, to be swiftly pulled out when the need takes you.**

Right above **Washing machines are always less noisy tucked behind doors.**

Right below **Food processors save time and energy, but make sure you keep all the attachments safely in one place to preserve your sanity.**

Far right **The expanse of stainless steel is relieved visually only by a fluorescent kettle on an otherwise empty surface.**

An oven, drawers, and appliance cupboards are all neatly housed in a stone cube in front of a supporting pillar where ample sockets allow for a display of useful but beautiful equipment.

Planning a kitchen around the existing architectural features of the room and incorporating contemporary units without compromising the space has been achieved here with success. One end of the room is devoted to functional, minimalist built-in cupboards that sit well beneath a cool marble work surface. A row of pigeonholes above the cupboards break up the solid expanse of white and provide a good display area for a collection of Grecian-style vases.

At the other end of the room, a more relaxed feel is created by the addition of a large wooden dining table, which incorporates a drawer for table linens, and other freestanding wooden units, which add warmth and soften the sophisticated modern designs elsewhere. A hefty butcher's block slots neatly into an unused fireplace where it makes an attractive feature that is useful for display. It has casters so it can easily be pulled out for use as a food preparation area, close to the dining table and adjacent to the well-stocked pantry where most of the essential ingredients live. A battered wooden hutch where the china is displayed stands against the opposite wall.

A place for everything

Instead of being used purely for display, this hutch is employed for storage, as it was originally intended to be. The white china looks great when it is not in use and is close to the table when meal times come around. Using a hutch in this way allows precious cupboard space to be devoted to other kitchen equipment that would look less attractive on display.

Left **The everyday china on
the shelves of a hutch
makes a simple display
between meal times.**

Right above **A generous
pantry cabinet is kept well
stocked with ingredients
for impromptu meals.**

This page **Unfussy white
built-in cupboards provide
a home for everything from
china to coats.**

Right below **Drawers inset
below a dining table house
table linens and flatware
where they are needed.**

Above **The pink of the cabinets is cleverly offset by the cool, natural shades of the walls, work surfaces, and floor, creating a clean, balanced, well-organized space, which is as functional as it is stylish.**

Left **A line of colorful built-in cupboards that matches the kitchen cabinets leads** the eye from one end of the house to the other and embraces all the clutter that no one wants to see.

Right **For an unusual but innovative solution, plates are stored neatly inside deep drawers that have been specially made with racks to hold them vertically in an ordered fashion.**

A converted basement is transformed into a stylish, contemporary kitchen that is made light and airy by the brilliant white walls, cool expanses of chrome and wooden counters, and sand-colored tiled floor, all of which lend the room a simple, understated atmosphere. In contrast are the bright, raspberry-pink cabinets with chic chrome pulls, which provide ample storage for all kitchen equipment and help to impose order and control in a busy home. The combination of colors and materials works well, as the natural tones of the wood, metal, and stone prevent the pink from being overpowering.

On the outside, at least, everything is organized, functional, and calm. The clutter is contained behind closed doors with only the essential appliances sitting ready in their place of use. Long counters provide plenty of space for preparing food, and the additional area on the central island, with the row of stools that tucks neatly underneath, is ideal for hasty breakfasts.

Cool and sleek with contemporary style

Tall cupboards that match the units are built all along one wall of an adjacent hall to tie the areas together visually and make light work of putting away toys, coats, and anything else that could disturb the symmetry of such a balanced, clean space.

Far left **Simple Cornishware china and a sink area below disguised with gingham fabric prove that a utilitarian kitchen can be very appealing.**

Top row left **Gilt china stored on industrial metal shelving makes a bold display all of its own.**

Bottom row left **Wooden flatware drawers are perfection, with everything neatly in place.**

Top row right
Mesh containers allow silverware to drain.

Bottom row right **Identical glasses in rows are both pleasing to look at and invitingly empty.**

Right above **Matching china stacked in neat piles always staves off a sense of panic when all around is chaos.**

Right below
An ergonomically designed flatware drawer allows simple access from one compartment to another.

I n a period house, the kitchen is designed to fit carefully among existing architectural elements so as to preserve their character while creating an unobtrusive, compact cooking area. The stove, like all good ranges, is slipped into the fireplace recess. A clock stands on the mantelpiece above as a hint of the room's former use. Next to the stove, the counter juts out slightly and is supported by tubular metal poles with shelves at intervals along their length, forming a stand for pots and pans right where you most need it. It also houses a serious collection of cooking oils, vinegars, and flavorings.

Tall cabinets with frosted-glass doors make full use of the high ceiling and provide an enticing but not wholly revealing view of the contents within. The upper shelves are filled with less-frequently-used items and are easily accessed by a pair of stylish foldout steps, which double as a stool. Built-in cabinets are divided into drawers and feel less formulaic because the wall cupboards are of a different, but sympathetic, design. All the essentials have been incorporated in a compact, ergonomic configuration in keeping with the existing features.

Compact, converted kitchen

Right **A pair of foldout steps are useful in all rooms of the home; in kitchens for reaching high shelves and cabinets, in children's rooms for reaching the top of closets to retrieve lost treasures, and they are essential for detaching drapes and shades in any room with tall windows.**

Center **A cooking range, counter, and glass-fronted cabinets slip happily in and around an old chimney recess to form a pleasing contemporary kitchen.**

Left **Wall-mounting a conventional plate rack and suspending other cooking implements from a rail saves a great deal of space.**

Below **A purpose-built pot stand juts out from the range but is close to hand for transferring pot to stove.**

Left **Knife storage has to be safe and practical. Creating slots within a kitchen drawer keeps individual pieces stationary, with the blades protected.**

Below **Wall-mounted knife storage is another good way of keeping them out of reach but ready for use.**

Right above **A spirited collection of colanders and jerry cans is brought to life with a garland of colorful flowers.**

Right below **Stainless-steel utensils in upright containers make shiny additions to the areas of kitchen action.**

Far right **Hanging racks needn't be elaborate. Inexpensive copper tubing is as good as shiny chrome poles for suspending equipment on butcher's hooks.**

The home office
_____
is here to stay

Work station

A working office is an increasingly common feature of many homes. The number of people who now spend all or part of their working week at home with their technology instead of commuting to the office is high, and rising all the time. Where once a humble study sufficed, consisting of no more than a solid oak desk tucked in a corner or under a window, many of today's new homes are being designed to include flexible spaces and rooms that are capable of incorporating a comprehensive home office.

Working from home can play havoc with your storage. Guest bedrooms, dining rooms, conservatories, landings, and under-stairs nooks are the spaces most often appropriated by home workers, only to be labeled too small, dark, public, and unprofessional by many lapsed office dwellers. The answer is economy. Economy of scale, belongings, filing, and equipment. Pare down the essentials to make your working life efficient and go from there. If your work involves a lot of paperwork that needs filing and has to be constantly accessible, you will need, at the very least, an alcove of shelves and a couple of cupboards to sustain all your storage needs. The natural cycle of work essentials, from note pads to stapler, letterheads to CD-ROMs, is one that involves storing them so you can find them, use them,

and keep them for the next time. A good mantra to employ when storing your work items is, "Find it, Use it, File it, Keep it." By having everything you may need contained in the immediate vicinity, your working chores will lessen, the ergonomics will function well, and your tasks will be completed quicker, making an early decamp to your other life a distinct possibility. Gauge just what furniture, technology, and materials you need to have close by, and you will be halfway to creating a successful working environment.

The simplest home office is often more akin to a traditional study area—somewhere quiet to pay bills, plan vacations, catch up on administration, and write letters. Such small-scale activities are easily accommodated in corners or halls, but once you work full-time or part-time from home and you need a desk, computer, fax, task lighting, and a book shelf, the demands increase. Self-contained, built-in units with neat pull-out drawers, fold-down counters and miniature screens and keyboards may be the answer if you are prepared to live a scrupulously compact working life. But unless you are one of those people who enjoy order at all costs, labeling every item in your possession and basking in the knowledge that you can locate anything within a minute, you may find yourself rebelling and spreading out belongings and reference material on the kitchen table at the thought of such restriction.

If you are working from home all the time, try to devote at least one room of the home to an office area. If you have space, it is worth having a self-contained place, which you can go to in the morning and leave at night. For locations, think original. Depending

Modern, flexible houses increasingly incorporate an ergonomic home-office space to suit a more fluid lifestyle.

Keep your work station separate from your leisure areas, whether it's on a landing, or in an attic or outhouse.

on your type of work, a room away from the house, in a converted garage or shed, may be more suitable. Craftspeople and artists, in particular, are often more at home in a studiolike environment, whereas more conventional office-based activities can quite happily be carried out in the home. Creating a platform in a room with high ceilings makes good use of space, as does a pull-down desk under the stairs or in an alcove. Under-used landings can be screened off and dedicated to a technology work station, or a shed can be transformed into an office away from the house. If you are a sea-

soned coffee drinker, you will need a coffeemaker within reasonable walking distance. Attic rooms are great for peace, but hardly practical if your front door and your kitchen where the cookies are kept are two floors down.

A dual-purpose room, where a workspace has to double as a guest bedroom or dining room, is possible as long as you demarcate the space and keep storage and furniture to a minimum. Sofa beds make good dividers, as do mobile cabinets and screens. Large armoire-like freestanding units are capable of housing all the technology and can have a pull-out or pull-down counter installed. Make sure all the work can be stored unobtrusively or preferably out of sight at the end of each day so your work doesn't impinge on your leisure too much.

Good lighting and seating are vital. Sacrifice some space if necessary in order to have a proper work chair. It will keep back problems at bay and add comfort to your working day. Preserve as

much natural light as possible—it will keep you cheerful when work problems overcome you, inspiration deserts you, or jobs dry up. Some say it is best not to position a work desk directly in front of a window as the view can be distracting. Others think it is a valuable diversion.

A work space is one room where the industrial sits happily with the traditional and function can dictate style. Metal filing cabinets and industrial shelving units on casters won't look out of place against a wooden map chest or glass-fronted oak cabinet. Decorate the space with an eye on the neutral; wood and metal are good companions, as are white and cream.

Let function dictate style in such an ordered environment, and keep your color scheme neutral.

Think about small-scale storage, too. Boxes, baskets, and mobile plastic or metal trolleys are all good solutions for paperwork, creative equipment, or a compact fax machine.

A vase of favorite flowers and other treats are called for

in an environment that lacks the constant social contact of an office. When you have to rely on yourself for company, good diversions and lunch-time breaks can be provided by music, the radio, an exercise bike, or the phone. You need a few distractions from the routine and a break from the solitude when you work at home.

Working away from the office is a burgeoning activity, so new homes are being designed with office space in mind and equipped with extra telephone sockets, plugs for computers, and natural light sources. Good storage is important. Depending on your work, you may need reference material and equipment, and a space for meeting clients. You will certainly find that a compact computer, comfortable chair and desk, uncluttered surface and plenty of shelves are vital if you need to house everything within a limited space. Make use of the entire floor-to-ceiling area to maximize your use of space. Shelves of varying heights will accommodate everything from magazines and files to oversized books and paperbacks, while small drawers, freestanding or incorporated in a larger system, hold stationery.

Include a radio or stereo if you are the kind of person who can only work with background noise. A small sink in a corner will be useful if your kitchen is far away, especially when you are visited by a client. Most important, make sure your chair and desk fit your requirements for comfort and ease of work.

Orderly

Left A window allows you a connection with the outside while you bash away at the computer. Have all you need close by and a chair on casters to save time.

Below A cup of coffee or glass of wine at the end of a busy day will set you up for the evening.

Right Adjustable shelving makes space for books of all sizes. Fill the top shelves with seldom-used books and keep a ladder handy.

Office in a corner

This page Loose baskets on open shelves hide paperwork, an old sewing table makes an informal desk, and a restored map chest provides storage space.

Opposite above The whole of one wall is devoted to work, with shelves and a sideboard with drawers.

Opposite below left **Ample natural light is cast over this desk, with a piano behind for diversion.**

Opposite below center **All the shallow drawers in this chest are cataloged for easy reference.**

Opposite below right **Fold-up chairs save space and are easy to store.**

Not all home offices have to be neat, structured affairs. Any corner can be given over to a small work space, thinking zone, or creative spot. One focal piece of freestanding furniture will set the tone, whether a map chest, a deep chest of drawers, or a simple table, you don't need much to get started. As long as there is enough storage for keeping paperwork under control, you can pay bills, write novels, rebuild a home, or start a business from a hutch or kitchen breakfast bar.

Walls are often the best spot for making space. Shallow shelves for keeping books or clutter; pigeon-holes for sorting fabrics, craft materials, or stationery; metal shelves for storing CDs and plants—all add something. If space is really tight, try a fold-up table

under a wall-mounted cupboard, or fit out a closet with a fold-down desk for an instant office. On an otherwise decorative piece of furniture, place a large wooden box to hold all your work paraphernalia. Use unexpected containers for unappealing junk. Baskets, metal trolleys, battered chests, and mobile filing cabinets all store and make a stylish statement.

Minimal office, maximum storage

Working at home involves clearing the mind of all things domestic, removing yourself from the humdrum of everyday life, and escaping to a space where the brain can function creatively and efficiently. Storage plays a strong role in creating the right kind of atmosphere in which to work successfully.

This minimal approach serves several functions. A simple desk with only the necessary elements at hand—computer, stationery, phone and fax, pencils and note pads—is hardworking yet not overwhelming. A light box for viewing work and large storage boxes for paperwork and other untidy clutter are arranged on movable, adjustable industrial shelving units that are infinitely flexible and don't fill the space. There are no visual distractions to disturb the concentration in this minimal area, and there is still room for the office to expand if need be.

Landings are good places for setting up small offices once sockets and phone lines are installed. Under-stairs cupboards can be taken out to create extra space and mobile storage units moved from room to room if needed. If you have a laptop or computer that lives on a mobile trolley, buy a desk which, when not in use, folds up into a small, stackable square to conserve space. Stacking boxes and files can be placed in mobile units and slotted into cabinets.

Left **Perspex™ folding chairs stack away at night and are unobtrusive additions to a minimal desk.**

Right **Sorted, stacked, and thoroughly cool, an office with a simple mission—efficiency.**

Creative space

Left Prevent the frustrated cry, "Has anyone see the scissors?" by attaching them to the work surface.

Right Numbered pigeon-holes are reminiscent of school lockers but seriously sensible for keeping bits and pieces in order.

Opposite above left
A pinboard, divided into sections with thick tape, makes a graphic framework for photographs and travel memorabilia.

Opposite above right
A galvanized pot attached to the wall makes a great holder for paintbrushes, with the added advantage that it cannot be knocked over.

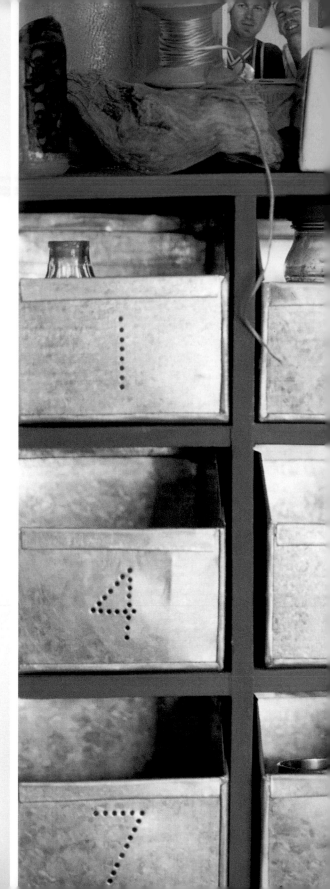

Below **Pens and pencils are
contained by a utilitarian
metal pot resting, in quirky
fashion, on a table mat.**

Whenever creative space is needed, there are usually disparate elements to draw together and keep safe but accessible. Use surprising containers to store everyday items and vice versa. Recycled cans look good en masse filled with pens or brushes; glass jars filled with buttons, braids, and shells present their own display; and fabric swatches glued to shoe boxes for easy identification are a good solution in a sewing room.

Whether a workroom is used for hobbies, to house the sewing machine and artwork, or as a home to a creative profession, like graphic design or costume-making, storing and displaying sources of inspiration becomes a strong part of the story. Displays of personal treasures may appear merely decorative, but will often spark off ideas. Leave space for inspirational items, and for cupboards and map chests in which to keep completed work in good condition.

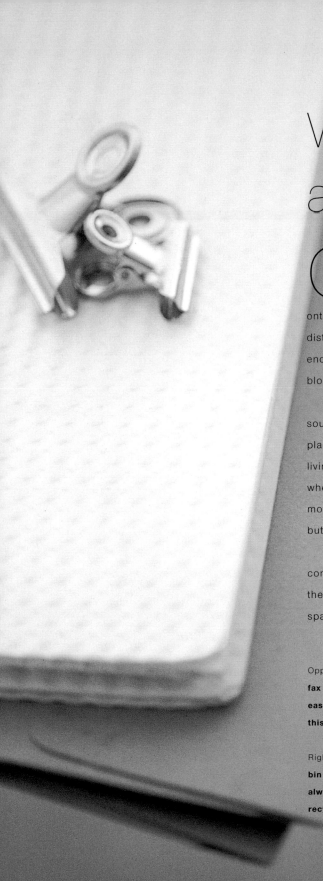

Working on a platform

Converted spaces often have tall, tall ceilings that need exploiting. This platform office lies halfway up the room, with a view onto the living space below. It would all be too distracting were it not for the metal-panelled enclosure, which brings containment and helps to block out noise from elsewhere in the home.

Compact offices can be more ergonomically sound than those in large spaces where everything is placed at random in a room. Organized, thought-out living is the only option in such a small space as this, where a work station stands next to a couple of mobile filing cabinets. That's all there is room for, but all that's required.

Cutting a worktop and shaping it to fit around the contours of a space, or else cutting it away to cross the corner of a room is a good way of making the space work for you.

Opposite **Computer, phone, fax and filing are all within easy reach of the desk in this compact space.**

Right top **Placing a waste bin next to the printer is always a good move – recycling made easy.**

Right center **Metal shelving reflects and extends the work ethic in an industrial setting.**

Right bottom **Small cardboard drawers are perfect for keeping stationery in order.**

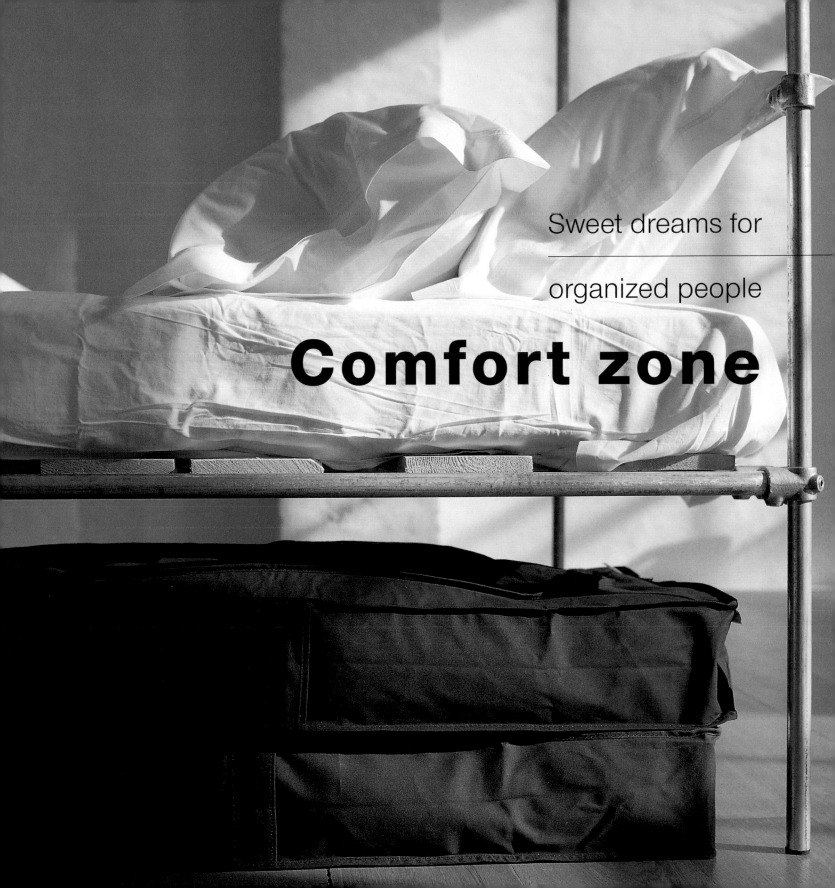

Sweet dreams for

organized people

Comfort zone

The bedroom is the one room in the home that is often dismissed as a place for sleeping and little else, yet it is the space where we spend a third of our lives. A sense of order and aesthetics is central to a quest for well-being. Sensitive architects, exponents of Feng Shui, and devotees of natural decorating are united in their belief that bedrooms containing natural materials, built-in storage, good lighting, and a sense of space are what we should be striving for to create somewhere restful, restrained, and reviving.

All sorts of disparate belongings find their way into the bedroom. Often it becomes a dumping ground for objects that can't be found space elsewhere, such as musical instruments, album collections, exercise equipment, or any books that overspill from the shelves in other rooms. In addition to this incidental clutter is the necessary task of providing space for clothes, shoes, accessories, toiletries, and bed linen.

The bed itself should be chosen and positioned with storage as well as a sense of style in mind. As the dominant feature in the room, it is what dictates the look and ethos of the space. Where tall ceilings prevail, a platform bed with a storage or work space beneath provides several solutions in one go. Beds can be as simple as slatted wooden units fitted into a recess with storage above and below, or as ornate as a wrought iron, antique extravaganza with calico drawers on casters underneath for hiding a shoe collection. Nowadays, under-bed storage is far more versatile and sophisticated than the old-fashioned notion of hording bank notes under the mattress. Purpose-made drawers, slimline baskets, and plastic boxes with lids and casters all provide space for blankets, sports equipment, and other items.

Once the bed is chosen and installed, you can exploit the area at the end of the bed. Blanket boxes, trunks, wooden shoe racks, and cubes on casters make good storage and display areas. Lying in bed contemplating a vase of favorite flowers, a sculpture, or a pile of treasured books at the end is far more enriching for the soul than gazing at a pile of discarded clothes, old toast, and yesterday's newspaper. Using the wall space immediately around the bed will often give further solutions. Building a simple unit on each side and above the head of the bed or sinking the bed itself into an alcove fitted with shelves are economic ways of using space, while having a bed base that sprouts bedside tables on each side as an organic extension to the bed frame is another. Think calm not chaos.

Much can be learned from the storage legacy of the Shakers, a religious movement of the 18th century based on the east coast of the United States. Their approach to life was simple, and every item in their possession was carefully stored in a place of its own, behind doors lovingly fashioned from wood, each joint, hinge, and spatial proportion carefully considered and executed. Their designs have become so sought after and influential because they make so much

> **Of all the rooms in a home, a bedroom should be an oasis of calm—a place where relaxation and rest come easily.**

The bed is the focal point of the room, and whatever the style, it will present all kinds of versatile storage options.

sense in a world where consuming is an obsession, possessions increase to the point of wastefulness, and the yearning for a simpler life is an inevitable ambition. The Shakers were among the first designers to resort to built-in cupboards for storing everything from honey to a week's worth of clothing and bed linen.

Tackling the storage of clothes is one of the main priorities in a bedroom. Start with the usual ruthless edit, create the piles to recycle or throw out, and think about how you are going to store what's left. Built-in closets are the best solution if you are planning to stay

somewhere for a long time. They conserve floor space and allow a good system of organization within. Grade all your clothes by size, then work out how much hanging space you need and at what intervals. Approach the insides of the closets as you would cabinets in a kitchen—consider everything you need to store, then buy or build appropriate elements. In the same way that carousel shelves, integral spice racks, and pot stands solve specific problems in the kitchen, closets can be outfitted with tie racks and drawers for scarves, socks, jewelry, and handkerchiefs, while shoes can be ordered in hanging racks or neatly lined up in pull-out drawers. Don't forget that clothes collections grow as well as shrink, so always allow room for expansion. Plan to install as many rails and shelves as you think you'll need, then add some. Allow space for awkward items, such as hat boxes, and tall slots for sports equipment. Look at fittings in stores for ideas and incorporate cardboard

shoes boxes, interior lights, plastic baskets, open shelves, and metal rails and boxes on casters as appropriate.

Calculate your hanging needs by allowing a space 25 inches wide for coats and dresses and 40 inches wide for jackets and shirts. But if all this measuring and calculation is one step too far along the road of self-organization, unfitted furniture could be a more appealing and flexible alternative. Folding or rolling T-shirts and tops are effective at keeping them crease-free; store

With freestanding rails and purpose-made racks and drawers, clothes storage has never been so exciting.

them in zip-up bags, boxes, or plastic cubes to keep the dust out. For a nomadic approach, pile sweaters and trousers into individual dump bins on casters and line them up under a freestanding clothes rail in an alcove. Conceal the area with a muslin drape and enjoy the

fact that you can be organized without resorting to total precision.

Old pieces of pleasing furniture, such as armoires, linen closets, chests of drawers and hutches can be adapted to hold clothes, shoes, boxes, and drawers, while open storage or a combination is visually appealing and versatile. Making the choice between unfitted furniture and built-in cupboards is often easier when faced with a room of irregular proportions, one that is small, or one that has to function as a dual-purpose space, where specific storage ideas will present themselves.

Above **The inside of this ordered closet is compartmentalized, with just the right amount of hanging space for jackets and trousers, shelves for laundered shirts, a deep drawer for sweaters, and a shelf for cuff links and belts, all concealed behind pristine white doors.**

Right **Many people's idea of heaven is a large walk-in closet or dressing room with seemingly endless**

space for storing clothes. Reminiscent of fittings in stores, this dressing area is fitted with rails for hanging dresses, skirts, shirts, and jackets, open pigeonhole sheves for sweaters and tops, which show exactly what's there, with an orderly rack at one end devoted to shoes.

Far right **It is essential to have plenty of good light in which to examine clothes before wearing them.**

Sorted, stacked, and hung

Clothes are never in one place for long. The journey from wearer to closet or washer and back again can be made easier by dedicating a specific space to the sorting and storage of clothes. It makes the ritual of choosing something to wear, checking it for marks and creases, wearing it, then washing and rehanging it an efficient process. A dressing area can be accommodated in a small space at one end of a room, or hidden behind the other side of a false wall, behind a headboard. Fitting it with pigeonholes for stacking tops, a variety of hanging rails at different heights for jackets, suits, and dresses, together with pull-out drawers for accessories and underwear means everything has its place. You can even go one step further and color code outfits.

This page **Scaffolding tubes** surrounding a bed give a certain sense of enclosure, and tie-on muslin drapes would provide privacy—the rails can even double as hanging space for clothes. In this minimal room, a ladder for reaching high storage spaces is used the rest of the time as a temporary book shelf, becoming a work of art in its own right.

Far right **Keep reading matter, photographs, and personal possessions in stacking boxes, so they are easily portable and calming to the eye.**

Right below **In a relaxed, informal space, storing a few aesthetically pleasing items, such as glossy books and soft blankets, becomes part of the decorative scheme.**

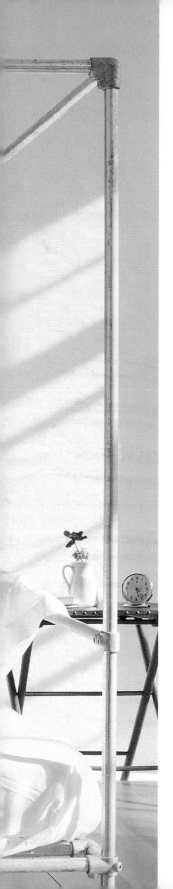

Freestanding and minimal

Beds slotted into frames such as scaffolding tubes, wooden four posters, salvaged hospital beds, and Shaker bedposts all exploit the potential for additional storage while making a graphic reference point in a bedroom. Add casters to the legs of the frame so you can move the bed around the room when the mood takes you. The antithesis of a built-in bed surrounded with fitted units and shelves, movable beds and furniture allow maximum flexibility and can help to accentuate a sense of space and light, especially in a modern room.

Storage in the form of under-bed boxes, bags, and metal trunks are as decorative as they are useful. Make sure any containers can be properly sealed so the contents don't become dusty. Store out-of-season shoes and seldom-worn clothes, bed linen, blankets, and sports equipment here, but avoid throwing reams of loose objects into the space; they will quickly gather dust and, knowing they're there,

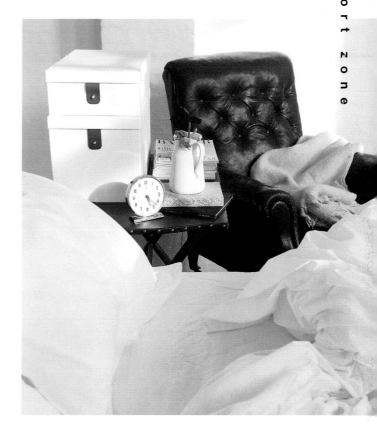

you will put off looking under the bed indefinitely. A couple of large, movable containers, though, are easy on the eye and are much more manageable to drag clear for cleaning under the bed.

Bedside tables, too, can be movable, foldable, and multipurpose. A fold-up table can easily be carried to another room when needed, or indulge yourself at the weekend and use it for eating breakfast. Small wooden, Perspex™, or plastic units on casters are all good alternative solutions for versatile bedside tables.

Left **Sometimes one plain, deep closet is all you need for your clothes, and this elegant piece can be moved around the house.**

Right **Thick, safelike doors swing open to reveal a carefully thought-out storage plan.**

Below **Swivel drawers conserve space and allow several to be opened at once if a search is on.**

Bottom **Shoes slot comfortably into the remaining spaces.**

The art of clever concealment

Once you have sifted through your clothes and discarded anything which is past its style date, worn, or faded, it may be worthwhile creating a one-off piece of furniture tailored to your needs. This ordered, elegant closet opens to reveal a seemingly limitless storage system based on simple clothes retrieval. Interior lights illuminate suits, jackets, and trousers, glass-fronted drawers glide open to reveal color-coded shirts, and ties are hung in a way that displays their pattern. Deep drawers hold sweaters, swivel-mounted boxes within the doors hold cuff links, glasses, and wallets. A generous mirror allows a final check on the inevitable elegance that follows effortlessly from such order.

Above and below **See-through containers are ideal for jewelry, hair accessories, or cotton balls.**

Right above **Pressed shirts are kept neat stored flat in individual drawers.**

Right below **The key to an organized closet is to group similar items and allocate them a space that is exclusively their own.**

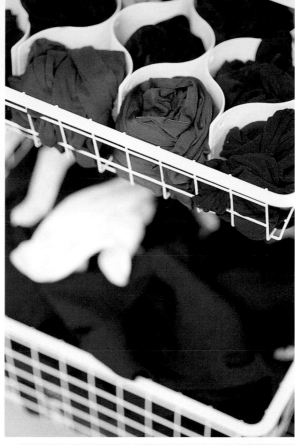

Top left **Pairs of shoes will remain together and in good condition on special racks.**

Top center **Drawer organizers will keep socks and pantyhose under control.**

Top right **A miniature chest of drawers is perfect for keeping jewelry in order.**

Far left **Tiny compartments store makeup so you can see exactly what's there.**

Left **Narrow shelves with doweling rims allow you to display rolled-up belts, suspenders, and scarves.**

This page **A versatile wall of storage is pure bliss. How could you not be organized with such a variety of drawers, rails, and shelves in such a glorious color?**

Right **A chic floor-to-ceiling storage system incorporates hanging space, lots of closets, and shoe storage slotted on the end. Neat drawer pulls punctuate the system in regular lines.**

Purpose-built solutions

Above **Generous baskets slot neatly under a metal bed to hide clutter and form a decorative, ordered line.**

Left **Baskets of bed linen nestle modestly behind a lace drape for a rural twist.**

Opposite top left **Boxes on top of cabinets provide additional storage space.**

Opposite top right **Fill up a battered old trunk with keepsakes, towels, bed linen, or blankets and place it at the end of the bed.**

Out of sight, out of mind

Left **A chest of drawers designed like fittings in stores is the perfect see-all clothes storage.**

f a strictly regimented built-in policy is simply not you, experiment with freestanding containers and furniture. As long as you have enough space to contain every last piece of clutter, a natural order will emerge. Freestanding pieces can be anything from blanket boxes and tin trunks to cane-paneled cabinets and store furniture. Any piece of furniture with shelves, compartments, drawers, or big doors is at once useful and decorative.

Baskets under beds are good for hair dryers and other ugly but frequently used items. Use them on shelves for bed linen, clean and dirty laundry, or children's toys. Retaining an element of visibility will help you identify the contents or at least jog your memory as to the last items you piled in.

Calmness, symmetry, and balance prevail in this beacon of modern ethnic design. Every surface, texture, and material has been carefully chosen and the furniture crafted in a spirit of creative control. The clever use of color ensures that the subtle, earthy shades merge unobtrusively from one to another. The walls incorporate flush-fitted cupboards of such finesse that they are enveloped by the space in a fusion of subdued colors. The bedside tables are useful, elegant cubes—even the closet door handles echo the rectangular shapes found everywhere.

Balanced space

Left **Designing storage that merges rather than emerges is a Modernist's dream.**

This page **A modest dressing area, where clothes and a chair don't hide the view, is neatly tucked behind a knocked-through wall.**

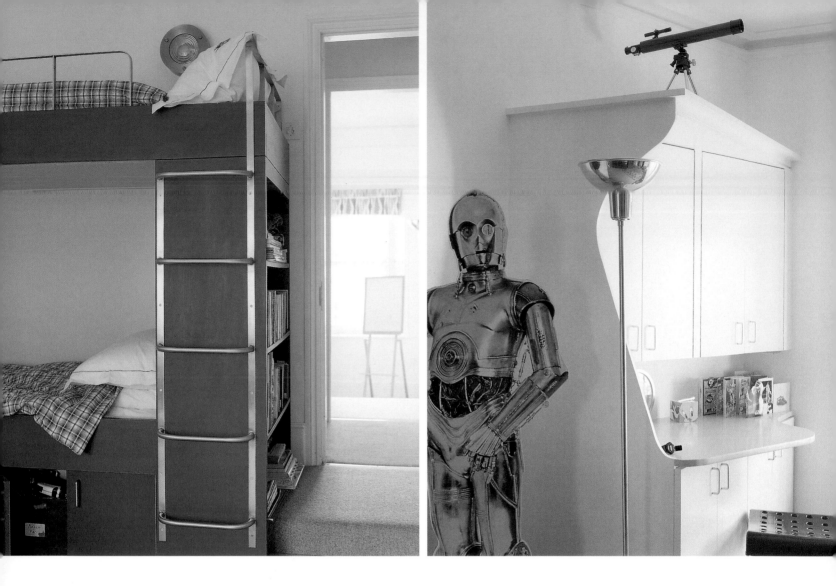

Storing the inevitable toys, clothes, and other objects that children gather in great, unwelcome quantities is always a challenge. Many parents see it either as an impossible task, or else they set off on a quest to impose order on everything from books to a mishmash of construction toys, board games, and incomplete jigsaw puzzles. Compartmentalized cabinets, boxes, drawers, and desks work well when children tidy up after themselves, take pride in their own space, and learn to love order. But you may never have time to devise the ultimate dolls closet or invent a way of scooping up the soft-toy hospital in one easy trawl. Take a deep breath, build lots of cabinets, and try the following.

Bunks and beds on platforms can often be built or adapted to include book shelves, work desks, lighting, and under-bed drawers. Equally, the dead space under a bed can be used effectively by filling it with baskets of toys or clothes drawers. Peg rails or wall-mounted canisters can be used to hold puppets, hair bands, drawstring bags of ballet gear, football equipment, and other items that are needed regularly. Boxes on casters are great for the end-of-the-day swoop to mop up soft toys, dressing-up clothes and other toys. Hanging clothes rails that can be wheeled about are useful for dressing-up clothes, school uniform, and party outfits, while hammocks are a good way of piling up soft toys.

Opposite left **A purpose-built bunk incorporates a ladder, safety rail, book shelves, and generous cabinets below.**

Opposite right **A corner study area combines storage and a contoured counter, which would be perfect for a computer.**

Left **Empty, decorative picture frames make handy, ornamental drawer pulls on a cubelike closet. Continuing the geometric theme, square wooden toy boxes on casters will please parents and children alike.**

Below **Colorful containers in funky shapes attached to the wall are both practical storage and display.**

Control and order in the playground

Clean, calm, revitalized,

and ready to go

Splash zone

Pure, clean, and simple, the bathroom is the place for pale colors, soft textures, and natural materials. Try to avoid, or at least minimize, bathroom clutter as much as possible, to create a tranquil setting that promotes a sense of calm for a total spiritual revival.

Once the ceramic fixtures are in place, bathroom storage is simply a matter of using any space that is left over in a way that both complements the style of room and ensures maximum efficiency for the grooming process, with optimum enjoyment when it comes to bathing. Much bathroom paraphernalia is less than wonderful to look at, so some form of concealment is usually necessary, whether in the form of neat built-in cupboards, less structured freestanding units, or informal, natural-colored muslin drapes. Solid materials, such as pastel or frosted glass, sleek beech wood, and clean, bright white tiles lend purity to cupboard doors, bathtub enclosures, and walls. Serenity beckons.

Many bathrooms have limited space, so cramming in piles of bath towels and bed linen, reams of toiletries, medicine, and first-aid supplies, and a laundry basket are often challenging storage exercises. As with kitchens, the smaller the space, the more likely you are to benefit from some kind of built-in unit, maybe one under the sink for storing all the cleaning materials or

a bath built into a recess that has shelves above and cupboards on each side. Safe storage is especially important in the bathroom, particularly for any sharp implements, toxic cleaning materials, and pills and medicines, which should always be kept in a lockable cabinet and out of the reach of children. Wall-mounted cabinets, either with or without mirrors, are ideal for storing medicines.

Nowadays, many toiletries are packaged in well-designed, visually pleasing bottles and jars that make a good display if they are left out when not in use. Narrow tongue-and-groove open shelves along one or more walls are the perfect way to store and display such items so you can always see exactly what's there.

In a small bathroom, you will feel instantly better about the lack of space if the towels are of good quality and a pleasing color; shampoo, bubble bath, and shower gel bottles look appealing rather than being plastered with a frenzy of advertising and labels; and soap, shaving equipment, toothpaste, and cotton balls are kept in attractive, interesting containers. Choose good-looking bars of soap and display them well in metal or wicker baskets, glass dishes, or metal trays.

When there is room in the bathroom for a chair, choose one with a lift-up seat that doubles as a laundry basket or a place to store clean towels, or clip a shelf onto the chair back for storing toys, body brushes, or toilet paper. In a small room, a folding chair or a stool that tucks under the sink or behind the door will conserve space. If the bathroom is large enough, incorporate a row of low-level cupboards along one wall for paraphernalia, and place bench cushions along the top. Hang a shelf close to the toilet for some

The bathroom should be more than a place to wash, it should be an oasis of calm where the spirits are revived.

Pare down the clutter and make sure you have room to keep luxurious treats to pamper yourself after a long day.

books and magazines. Glass shelves are always pleasing in a bathroom, since the light that bounces off mirrors, faucets, and gleaming sinks is reflected and enhanced as it hits glass surfaces.

Alternatively, dispense with shelves and install a large, long peg rail instead. Two pegs for each member of the family—one for towels and one for a drawstring bag of dirty laundry or personal toiletries. Use the back of the door for coat hooks so robes can keep dry. Store towels either on traditional freestanding towel rods, on wall-mounted and heated chrome rods, or on pivoting rods mounted on the wall. Make sure they are within easy reach of the sink, shower, or bath to be ergonomically sound. Freestanding furniture such as linen cupboards, or movable cupboards and trolleys, are also good ways of storing spare towels, clean bed linen, and extra blankets from the bedroom.

In showers, a neat wall-hung metal shower shelf will hold all the shampoos, conditioners, and soaps you are likely to need. Clip additional plastic hanging rails over a shower door to hold a towel, slippers, and glasses. Bath racks placed over the tub are a good way of keeping everything you need for a relaxing soak on hand.

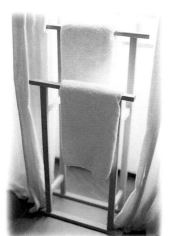

Supplement with a book holder, a glass of wine, and a scented candle for relaxing nighttime indulgence.

The bathroom is often a good place for a dressing table and full-length mirror (as long as you don't mind the scale having the same home). Sometimes it makes sense for bathrooms to double up as utility rooms. It is not difficult to adapt bathroom plumbing to include a washing machine and tumble dryer, in which case the natural life cycle of dirty clothes leaving a person, going into the laundry basket, and thence to the washing machine is considerably speeded up. Add a ceiling-mounted hanging rack over the bathtub for clothes to dry on, and a couple of hours of spare time beckon. This also means that all

laundry is neatly contained in one area rather than allowing it to spread over several rooms. Washing machines can be stacked on top of dryers to conserve space and hidden behind cupboard doors, which may also incorporate a fold-down ironing board and other cleaning materials, as well as additional laundry baskets for color coding the wash or to accommodate the clean, dry "to iron" pile. Fit large laundry baskets with lift-out baskets with handles to make transporting dirty clothes to the washing machine quicker and easier. Then return the inner basket to the utility area full of

Make a display of pleasing bottles and soaps, and use soft, natural-fiber towels in soothing colors.

clothes that are ready to iron. Alternatively, use a stylish, modern fabric laundry bag on a metal or wooden frame on casters for wheeling easily from room to room to collect the dirty clothes.

The bathroom is an obvious place for displaying a collection of water-related treasures. Fossils, model boats, and pretty shells make wonderful seaside still lifes on shelves or other surfaces, while worn glass fragments and stones piled into glass tanks evoke a sense of the coast and allow us a moment's mental transportation. Protect delicate wooden surfaces with a few layers of matte varnish to prevent condensation or splashes from causing damage.

Left **The open shelf under the built-in sink unit breaks up the expanse of wood and prevents it from dominating.**

Right **Colored and natural baskets in sisal and wicker scoop up all the usual bathroom paraphernalia and provide decoration with their repeating patterns.**

Opposite **Ultra-sleek wall-mounted taps and a stylish glass sink, which rests unobtrusively on a built-in unit, conserve space for all-important grooming.**

The art of control and display

The art of control is nowhere better exemplified than in this bathroom, where everything is neatly stored but still visible and accessible. Such a keen sense of order instills purpose and energy at the beginning of the day and a pervading aura of calm at night. The storage is a combination of built-in and stand-alone units that are all perfectly functional and provide a natural home for everything. The array of baskets house difficult-to-store items in sensible, contained order, including rolled-up pastel towels that are easy to grab hold of, and add color and texture to the otherwise-neutral decor. A recess between the door and unusual built-in sink unit is put to good use with floor-to-ceiling pigeonhole shelves.

Create visual peace and inner harmony in a neutral bathroom by choosing easy containers in glass, plastic, or delicate porcelain and scrupulously white bath linens. A room that is clean, white, and airy, with good lighting and a firm control of space promotes peace, and carefully planned storage helps to maintain the sense of calm. Keep towels, toiletries, accessories, and medicines under control in hidden spaces or on discreet display to allow the space around you to breathe, soothe, and heal.

Bathing is one of life's repeated pleasures, a constant cycle of renewal, which revives and restores the balance of a busy live. Organizing a bathroom so you can wash, groom, bathe, shower, dry, and preen means that the storage must work for you as much as the space does. Unsightly cleaning equipment should be kept out of sight, while toiletries and towels should be easy to reach and appealing.

Pure, clean, and simple

Far left Soaps are stored in simple porcelain soap dishes.

Far left center Unobtrusive containers in different sizes have understated style lined up on a shelf.

Far left bottom A clean bathmat is at the ready draped over the edge of the bathtub.

This page A clutter-free bathroom is the epitome of tranquility; for creating a sense of calm and restoring peace of mind, there is nothing better than the combination of natural wood and pure white.

Top right Pristine towels are piled up next to the basin, right where they're needed.

Bottom right A laundry basket is kept in the hall, on route from the bathroom to the utility room.

This bold homage to high tech incorporates a
grooming area, clothes storage, and bathroom
in a large, light-filled space. Strictly industrial
materials, such as rubber flooring, a Modernist
pedestal sink unit, and frosted sliding panels merge
happily with a contemporary take on a 1930s dressing
table and two seaside-inspired semicircular canvas
screens that offer privacy without obscuring the light.
One canvas cubicle encloses a freestanding bathtub

Sleek, stylish industrial grooming

at one end of the space; another at the opposite end
conceals a floor-draining shower. Along one wall of
the space, clothes are kept behind the doors, always
on hand to slip into once grooming is completed.

Far left **Phillipe Starck** drawer handles lend urban modernity to a glass, wood, and steel dressing table that rests between the two canvas cubicles as a neat space divider.

Left **Flexible storage in** the form of hanging rails, shelves, and drawers are discreetly stowed behind semiopaque sliding doors, which also hide the toilet.

Above **Chrome towel rods** are integral to the pedestal sink units and swivel conveniently for use.

Right **Folding steps** allow access to the upper reaches of the closet. The sliding doors can be left open for clothes to be selected from the comfort of the bathtub.

Far left The most humble contents can be given graphic resonance by the right container.

Center top left A dressing table incorporates storage for grooming equipment, while a miniature plastic chest of drawers holds jewelry and accessories.

Center top right Freestanding chrome trays, with handles for easy portability, are ideal for storing well-packaged toiletries.

Center bottom left The bath tray: simple, useful, and elegant over a gleaming white bathtub, stacked with pleasing bathtime treats.

Center bottom right Discarded medical equipment eases the pressure on toothbrush storage.

Right above Unexpected containers often double as display. Multicolored nylon shopping bags house clean towels and transport dirty ones to the laundry basket.

Right below Soaps in multicultural wrappers are far too gorgeous to store in a cupboard; they need a home where they can be seen.

Far left **Bathing becomes an ethereal experience in a bathroom divided from the bedroom by double doors, which open to allow light to dance over all the surfaces.**

Left **Natty Perspex™ cubes serve the dual function of protecting fragile fossils and providing a perfect showcase for the collection.**

Right **Towels, clothes, and art mingle in a room where display and storage are one and the same.**

Showcase solutions

Sometimes the bathroom is the only logical place for housing a personal collection of water-related treasures. In this big, white space, large windows cast natural light and an aura of calm onto a stunning collection of fragile, translucent fossils placed within a purpose-built arrangement of wooden pigeonholes that make a focal point of the area above the sink. The unit is set against a mirrored alcove and protrudes slightly from the wall to accommodate square Perspex™ cubes, each one housing a different, delicate cargo. The light that comes into the room bounces back and forth between the wall and mirror in a cheering manner. Sandstone that covers the bathtub and sink provides a textural reference to the collection and, in turn, is home to bowls of large shells and water-smoothed pebbles. Making a virtue and a feature of your collections and belongings in this manner means they are always accessible and available to enjoy, and at the same time make a room stylishly unusual.

Bed and table linen, towels, and clothes are never static for long. Unlike a jewelry box or tie collection that stays in one place, they follow a cycle as they move from dirty luandry basket to washer, dryer, ironing basket, then back to storage before the process begins again. Storing such frequently moving items so they progress through all the stages with a minimum of pile-building, sock-losing, and ironing overload is almost a scientific task. As is often true, lots of storage can work wonders. Helpful solutions are to have baskets, maybe with casters, in the bedroom, by the washer, and wherever the ironing is done, and plenty of hangers on hand.

Ironing out those creases

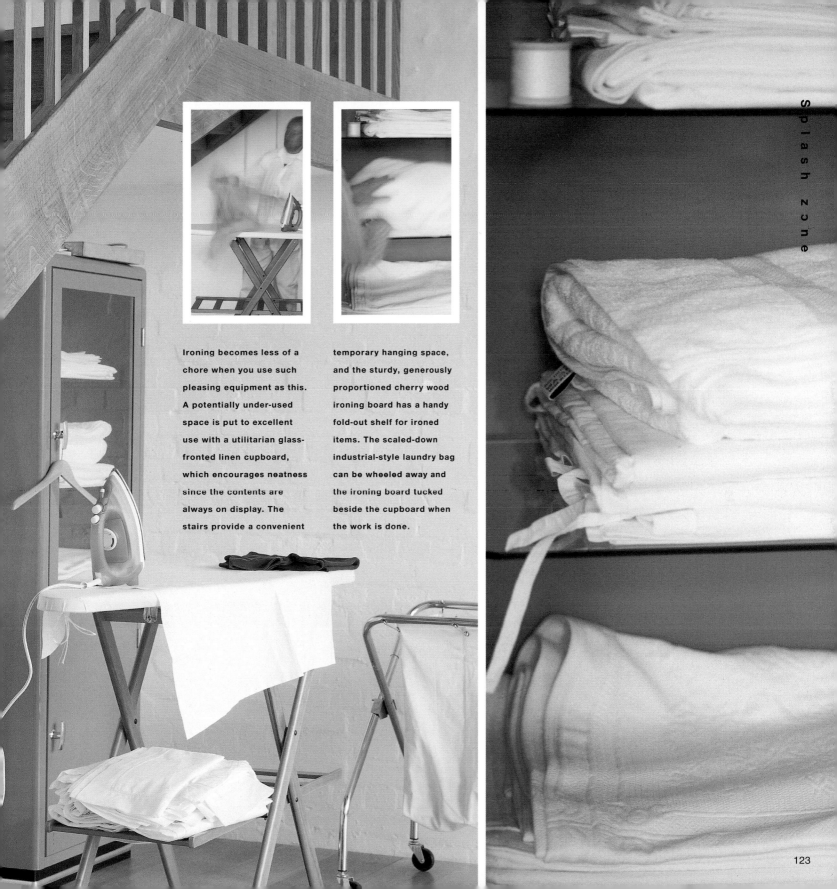

Ironing becomes less of a chore when you use such pleasing equipment as this. A potentially under-used space is put to excellent use with a utilitarian glass-fronted linen cupboard, which encourages neatness since the contents are always on display. The stairs provide a convenient temporary hanging space, and the sturdy, generously proportioned cherry wood ironing board has a handy fold-out shelf for ironed items. The scaled-down industrial-style laundry bag can be wheeled away and the ironing board tucked beside the cupboard when the work is done.

Source directory

The publisher and the authors of *Storage* are not responsible for the products sold by the following companies, and it is not our intention to promote any of these purveyors.

Kitchens

Christians
212-333-8794
Now widely available in the U.S., this British manufacturer offers the best in fitted furniture for the kitchen in five classic styles.

Downsview Kitchens
2635 Rena Road
Mississaugua
Ontario
Canada L4T JG6
Fax: (416) 677 5776
Available throughout the U.S.

Holiday Kitchens
A division of Mastercraft Industries, Inc.
120 West Allen Street
Rice Lake, WI 54868
715-234-8111
Fine custom kitchens at excellent prices have kept this company in business for over 50 years. Cabinets are shipped all over the country.

Ikea
To obtain catalog, write:
Ikea Catalog Department
185 Discovery Drive
Colmar, PA 18915
Offers a wide range of stylish, innovative, and affordable cabinetry designs that can be customized to your needs. Design service, delivery, and installation available.

Plain & Fancy Custom Cabinetry
P.O. Box 519
Schaefferstown, PA 17088
717-949-2924
The New American classics. Custom cabinetry at an affordable price.

Poppenpohl U.S., Inc.
145 U.S. Highway 46 West
Suite 200
Wayne, NJ 07470
800-987-0553
Clean, modern lines are the signature of this firm. They offer more than 180 front styles in wood, lacquer, and laminates.

Quaker Maid Kitchens
140 Commerce Drive
Rochester, NY 14623
800-992-56-24
Specializes in combining standard components with care to create custom-look kitchens.

Sears, Roebuck and Company
800-469-4663
Working with local contractors, Sears resurfaces cabinets and drawers to change the look of a kitchen at minimum cost.

SieMatic Corporation
P.O. Box 936
Langhorne, PA 19047
(800) 765 5266
Kitchen interior design, with modular units for flexibility and style.

Wellborn Cabinet, Inc.
P.O. Box 1210
Ashland, AL 36251
800-762-4475
Offers stock wooden cabinets in many different sizes with many different fronts, giving thousands of options for a seemingly custom-designed kitchen.

Wood-Mode Custom Cabinetry
Kreamer, PA 17833
800-635-7500
This extensive line features many door styles, finishes, and special-purpose built-in features.

Furniture and Storage Units

ABC Carpet & Home
888 Broadway
New York, NY 10003
212-473-3000
A great selection of furniture.

Bed, Bath & Beyond
620 Sixth Avenue
New York, NY 10011
516-424-1070
Everything for the bedroom and bathroom and the kitchen too.

Berkeley Design Shop
2970 Adeline Street
Berkeley, CA 94703
510-841-5340
Furniture and accessories.

Calvin Klein Home
Flagship store:
654 Madison Avenue
New York, NY 10022
212-292-9000
00-294-7978 for a store near you.

Chambers
P.O. Box 7841
San Francisco, CA 94120
(mail order)
Everything for the bedroom and bathroom.

Closet Maid
720 West 17th Street
Ocala, FL 34478
800-221-0641
Furniture and cabinetry.

Colonial Williamsburg Foundation
Department 023
P.O. Box 3532
Williamsburg, VA 23187-3532
800-446-9240
Colonial-style home accessories.

Crate and Barrel
650 Madison Avenue
New York, NY 10022
or
Crate and Barrel
P.O. Box 9059
Wheeling, IL 60090-9059
800-451-8217
Good-value furniture & accessories.

Details at Home
1031 Lincoln Road
Miami Beach, FL 33139
305-531-1325
All kinds of accessories.

Equipto
225 South Highland Avenue
Aurora, IL 60507
800-323-0801
Good basic storage units.

Evergreen Antiques
1249 Third Avenue
New York, NY 10021
212-744-5664
Armoires and closets.

Gracious Home
1220 Third Avenue
New York, NY 10021
212-517-6300 or 800-338-7809
Everything you need for all rooms.

Hope & Wilder
454 Broome Steet
New York, NY 10013
212-966-9010
Decorative old furniture.

La Barge
Dept H90
P.O. Box 6917
Holland, MI 49422
616-392-4666
Old world pine finishes, woven
rush seating, and glazed
ceramic furnishings.

Laura Ashley, Inc.
6 St. James Avenue
Boston, MA 02116
800-367-2000
For a catalog call 800-429-7678.
English-style closets and hutches.

Organized Designs
P.O. Box 6901
Beverley Hills, CA 90212
310-277-0499
Cabinets and closets.

Palecek
P.O. Box 225
Richmond, CA 94808
800-274-7730
Manufacturers of fine crafted wicker,
rattan, and wooden accent furniture.
Call for a dealer in your area.

Pier 1 Imports
For a store in your area call
800-44PIER1 (800-447-4371)
All kinds of stylish accessories
for all the rooms in the home.
Nationwide locations.

Pottery Barn
P.O. Box 7044
San Francisco, CA 94120-7044
800-588-6250
Home accessories. Stores Nationwide.

Ralph Lauren Home Collection
1185 Sixth Avenue
New York, NY 10036
For a store in your area, call
800-377-POLO (800-377-7656)
Stylish accessories.

Roche Bobois
183 Madison Avenue
New York, NY 10016
212-889-5304
Furniture and storage from Paris.

Ruby Beets
Poxabogue Road and Route 27
P.O. Box 596
Wainscott, NY 11932
516-537-2802
Vintage painted furniture, old white
china, and kitchenware.

Staples
1075 Sixth Avenue
New York, NY 10018
800-333-3330
Cabinets and furniture.

Wolfman Gold & Good, Inc.
117 Mercer Street
New York, NY 10012
212-431-1888
Furniture and accessories—some
old, some new, all delightful.

Zona
97 Green Street
New York, NY 10012
212-925-6750
Home environment shop with a
rustic, southwestern point of view.

Small-scale Storage

Hold Everything
P.O. Box 7807
San Francisco, CA 94120
or
Hold Everything
865 Market Street
San Francisco, CA 94103
415-546-0986
Everything for storage from baskets
to shoe holders. Mail order too.

Workspaces

California Closet Co.
3385 Robertson Place
Los Angeles, CA 90034
818-705-2300
1-800-336-9206 for more information.
Custom-designed closets to meet all
storage needs for the home office
and more.

Sam Flax
12 West 20th Street
New York, NY 10011
212-620-3000
Industrial cabinets.

Acknowledgments

1 a house in London designed by Guy Stansfeld 0171-727 0133; **2** Dawna & Jerry Walter's house in London; **3** *left* Paula Pryke's house in London; **3** *right* Vanessa & Robert Fairer's studio in London designed by Woolf Architects 0171-428 9500; **4** *center* the London flat of Miles Johnson & Frank Ronan; **4** *right* Vanessa & Robert Fairer's studio in London designed by Woolf Architects 0171-428 9500; **5** Rosa Dean & Ed Baden-Powell's apartment in London, designed by Urban Salon 0171-357 8800; **6** *above* Vanessa & Robert Fairer's studio in London designed by Woolf Architects 0171-428 9500; **6–7** Rosa Dean & Ed Baden-Powell's apartment in London, designed by Urban Salon 0171-357 8800; **8–9** Robert Kimsey's apartment in London designed by Gavin Jackson 0705 0097561; **9** the London flat of Miles Johnson & Frank Ronan; **10** *left* Paula Pryke's house in London; **10** *above right* Paula Pryke's house in London; **10** *center* a house in London designed by François Gilles and Dominique Lubar, IPL Interiors; **10** *below* The London flat of Miles Johnson & Frank Ronan; **11** *center left* Ian Bartlett & Christine Walsh's house in London; **11** *below left* Paula Pryke's house in London; **11** *right* Vanessa & Robert Fairer's studio in London designed by Woolf Architects 0171-428 9500; **12–13** Architecture & furniture by Spencer Fung Architects tel/fax 0181-960 9883; **13** *above left* Paula Pryke's house in London; **13** *center left* Dawna & Jerry Walter's house in London; **13** *below left* Vanessa & Robert Fairer's studio in London designed by Woolf Architects 0171-428 9500; **14** *left* a house in London designed by Guy Stansfeld 0171-727 0133; **15** *above left* a house in London designed by François Gilles and Dominique Lubar, IPL Interiors; **15** *right* Architecture & furniture by Spencer Fung Architects tel/fax 0181-960 9883; **16–17** David Jermyn's house in London, designed by Woolf Architects 0171-428 9500; **18** Robert Kimsey's apartment in London designed by Gavin Jackson 0705 0097561; **19** *left* Robert Kimsey's apartment in London designed by Gavin Jackson 0705 0097561; **19** *right* Vanessa & Robert Fairer's studio in London designed by Woolf Architects 0171-428 9500; **20–23** David Jermyn's house in London, designed by Woolf Architects 0171-428 9500; **23** Robert Kimsey's apartment in London designed by Gavin Jackson 0705 0097561; **24** Rosa Dean & Ed Baden-Powell's apartment in London, designed by Urban Salon 0171-357 8800; **25** Architecture & furniture by Spencer Fung Architects tel/fax 0181-960 9883; **27** *left* Rosa Dean & Ed Baden-Powell's apartment in London, designed by Urban Salon 0171-357 8800; **28–29** Robert Kimsey's apartment in London designed by Gavin Jackson 0705 0097561; **29** *above left* Paula Pryke's house in London; **29** *below left* David Jermyn's house in London, designed by Woolf Architects 0171-428 9500; **29** *right* Robert Kimsey's apartment in London designed by Gavin Jackson 0705 0097561; **30–31** A house in London designed by Guy Stansfeld 0171-727 0133; **32–33** Paula Pryke's house in London; **34** Vanessa & Robert Fairer's studio in London designed by Woolf Architects 0171-428 9500; **34–35** Dawna & Jerry Walter's house in London; **35** *left* Vanessa & Robert Fairer's studio in London designed by Woolf Architects 0171-428 9500; **35** *right* David Jermyn's house in London, designed by Woolf Architects 0171-428 9500; **36–39** Architecture & furniture by Spencer Fung Architects tel/fax 0181-960 9883; **40** *left* Dawna & Jerry Walter's house in London; **40** *center* Rosa Dean & Ed Baden-Powell's apartment in London, designed by Urban Salon 0171-357 8800; **40–43** Rosa Dean & Ed Baden-Powell's apartment in London, designed by Urban Salon 0171-357 8800; **44** *left* the London flat of Miles Johnson & Frank Ronan; **44** *above right* Vanessa & Robert Fairer's studio in London designed by Woolf Architects 0171-428 9500; **44** *below right* Rosa Dean & Ed Baden-Powell's apartment in London, designed by Urban Salon 0171-357 8800; **45** Vanessa & Robert Fairer's studio in London designed by Woolf Architects 0171-428 9500; **46**

Dawna & Jerry Walter's house in London; **49** *below* Dawna & Jerry Walter's house in London; **52–53** Dawna & Jerry Walter's house in London; **56** *below right* Dawna & Jerry Walter's house in London; **57** *above right* Ian Bartlett & Christine Walsh's house in London; **62** *above left* Vanessa & Robert Fairer's studio in London designed by Woolf Architects 0171-428 9500; **62** *right* Robert Kimsey's apartment in London designed by Gavin Jackson 0705 0097561; **63** *above left* Paula Pryke's house in London; **63** *right* Paula Pryke's house in London; **66–67** a house in London designed by Guy Stansfeld 0171-727 0133; **68** *left* the London flat of Miles Johnson & Frank Ronan; **68** *above right* Paula Pryke's house in London; **68** *below right* the London flat of Miles Johnson & Frank Ronan; **69** *above right* Dawna & Jerry Walter's house in London; **70–71** Ian Bartlett & Christine Walsh's house in London; **73** *above left* the London flat of Miles Johnson & Frank Ronan; **73** *right* Ian Bartlett & Christine Walsh's house in London; **75** a house in London designed by François Gilles and Dominique Lubar, IPL Interiors; **76** *above right*

a house in London designed by François Gilles and Dominique Lubar, IPL Interiors; **77** *left* the London flat of Miles Johnson & Frank Ronan; **77** David Jermyn's house in London, designed by Woolf Architects 0171-428 9500; **78–79** Ian Bartlett & Christine Walsh's house in London; **81** *above* the London flat of Miles Johnson & Frank Ronan; **81** *below left* a house in London designed by Guy Stansfeld 0171-727 0133; **84** *left* the London flat of Miles Johnson & Frank Ronan; **84–85** the London flat of Miles Johnson & Frank Ronan; **88** a house in London designed by François Gilles and Dominique Lubar, IPL Interiors; **89** Vanessa & Robert Fairer's studio in London designed by Woolf Architects 0171-428 9500; **90** *above left & below* Paula Pryke's house in London; **90** *above right* Dawna & Jerry Walter's house in London; **91** *left* Paula Pryke's house in London; **91** *above right* a house in London designed by François Gilles and Dominique Lubar, IPL Interiors; **91** *below right* a house in London designed by François Gilles and Dominique Lubar, IPL Interiors; **92–93** a house in London designed by

François Gilles and Dominique Lubar, IPL Interiors; **94–95** Vanessa & Robert Fairer's studio in London designed by Woolf Architects 0171-428 9500; **96–97** a house in London designed by François Gilles and Dominique Lubar, IPL Interiors; **98** *left* Paula Pryke's house in London; **98** *right* a house in London designed by François Gilles and Dominique Lubar, IPL Interiors; **99** *above left* a house in London designed by François Gilles and Dominique Lubar, IPL Interiors; **99** *above center, above right & below* Paula Pryke's house in London; **100** Ian Bartlett & Christine Walsh's house in London; **101** a house in London designed by Guy Stansfeld 0171-727 0133; **102** *above* Dawna & Jerry Walter's house in London; **102–103** Architecture & furniture by Spencer Fung Architects tel/fax 0181-960 9883; **103** *left* Dawna & Jerry Walter's house in London; **103** *right* the London flat of Miles Johnson & Frank Ronan; **104–105** Architecture & furniture by Spencer Fung Architects tel/fax 0181-960 9883; **106** a house in London designed by Guy Stansfeld 0171-727 0133; **109** Robert Kimsey's apartment in London

designed by Gavin Jackson 0705 0097561; **110** *above right* Paula Pryke's house in London; **110** *below* Vanessa & Robert Fairer's studio in London designed by Woolf Architects 0171-428 9500; **111** a house in London designed by François Gilles and Dominique Lubar, IPL Interiors; **112–113** Dawna & Jerry Walter's house in London; **116–117** Paula Pryke's house in London; **118** *left & above right* Paula Pryke's house in London; **118** *below right* the London flat of Miles Johnson & Frank Ronan; **119** *above left* Paula Pryke's house in London; **119** *below right* the London flat of Miles Johnson & Frank Ronan; **120–121** a house in London designed by Guy Stansfeld 0171-727 0133; **122–123** Vanessa & Robert Fairer's studio in London designed by Woolf Architects 0171-428 9500; **128** Dawna & Jerry Walter's house in London.

The publishers would like to thank the following companies for loaning props for photography for this book: Grand Illusions The Holding Company McCord Storage Paperchase

Authors' acknowledgments

We would like to thank the multi-talented Andrew Wood for his
funky photography and good company and the marvelous Melanie Molesworth
for her effortless style. Special thanks to Nadine Bazar and Karina Garrick,
and warm thanks to the wonderful team at RPS: Jacqui Small, Anne Ryland,
Kate Brunt, Zia Mattocks, and Ashley Western.

Thanks to our photographic models: Kate Brunt, Becky Davies,
Elizabeth of Mar, Zia Mattocks, Melanie Molesworth, Ashley Western, and
Andrew Wood. And thanks to Melanie Molesworth's assistant, Serena Hanbury.

Many thanks also to the people who allowed us to photograph their homes.